IRRIGATION CANAL LINING

FAO Land and Water Development Series   No. 1

IRRIGATION CANAL LINING

by

D.B. KRAATZ

Hydraulic Engineer
FAO Land and Water Development Division

FOOD AND AGRICULTURE ORGANIZATION OF THE UNITED NATIONS
ROME, 1977

The designations employed and the presentation of material in this publication do not imply the expression of any opinion whatsoever by the Food and Agriculture Organization of the United Nations concerning the legal status of any country, territory, city or area or of its authorities, or concerning the delimitation of its frontiers or boundaries. The views expressed are those of the author.

P-56
ISBN 92-5-100165-0

The copyright in this book is vested in the Food and Agriculture Organization of the United Nations. The book may not be reproduced, in whole or in part, by any method or process, without written permission from the copyright holder. Applications for such permission, with a statement of the purpose and extent of the reproduction desired, should be addressed to the Director, Publications Division, Food and Agriculture Organization of the United Nations, Via delle Terme di Caracalla, 00100 Rome, Italy.

© FAO 1977

PREFACE

The purpose of this manual is to provide a compilation of the available information on the techniques of irrigation canal lining. The lining of small canals and ditches has been emphasized because, despite their great importance in the efficient use of irrigation water, they are often neglected by project planners and engineers. The manual should be useful for planners, design engineers, technicians, extension services and field project staff, and it is hoped that it will stimulate the exchange between various geographical areas of lining techniques and designs which have evolved independently.

Acknowledgements. The author wishes to express his gratitude to all those individuals, organizations and firms that contributed to this manual by providing valuable information, illustrations and constructive criticism and comments.

Special acknowledgement is made to C.E. Houston, Chief, Water Resources Development and Management Service, FAO, and to the late C.W. Lauritzen, who reviewed this manual. Mr. Lauritzen was an outstanding authority on irrigation engineering and frequently gave advice and assistance to FAO.

Notes. In this manual measures and weights are generally expressed in metric terms followed in brackets by the equivalent in the English system. Exceptions are those instances in which it seemed preferable to retain the system used in the source, usually reprints of tables, figures and direct quotations. The term "cusecs" is sometimes used for cubic feet per second. Dollars ($) always refers to U.S. dollars.

Bibliographic references are divided into descriptive categories lettered A, B, C, D, E, and F, under which the entries are numbered, in sequence. Thus, references in the text are cited, for example, as A1, B4, B15.

CREDITS FOR ILLUSTRATIONS

The following individuals, bibliographical sources, enterprises, and public institutions are gratefully acknowledged for illustrations reproduced in this publication:

American Society of Agricultural Engineers (ASAE): Figs 27, 29-31; Binnie & Partners, Chartered Civil Engineers, London: Figs. 84, 85; H. Bower (B10): Figs. 6-10; B.W. Culy (FAO): Fig. 48; Dula-Navarrete (FAO): Fig. 46; Fullerform Inc., Phoenix, Arizona, U.S.A.: Fig. 56; Gorokrik (C34): Fig. 64; R.A. Hanson Co., Inc., Spokane, Washington, U.S.A.: Fig. 57; F.H. Herrero (A3): Figs. 17-18; C.W. Lauritzen (C58): Figs. 2, 3, 19, 42-44, 49-51; (C60): Figs. 97, 98; C.W. Lauritzen & R.E. Griffin (C61): Figs. 52, 53; L. Lovas (FAO): Figs. 33, 78 a, b, 90, 91; D. Mason (FAO): Fig. 4; A.M. McIntyre (A3): Figs. 47, 54, 55, 62; K. Pohjakas (FAO): Figs. 39-41; Portland Cement Association (PCA), Chicago, Illinois, U.S.A.: Figs. 23, 28, 45, 60, 61, 63, 96; J. Pourtauborde (FAO): Figs. 67, 68; Punjab, Irrigation and Power Department (D10): Figs. 86, 87; K. Roscher (FAO): Figs. 20, 24; J.A. Sagardoy-Alonso (FAO): Fig. 5; R. Sentenac (FAO): Figs. 32, 75; Spain, Ministry of Public Works (M.S. Barragan): Figs. 58, 59; C.E. Staff (C75): Fig. 83; U.S. Department of Agriculture, Agricultural Research Service: Fig. 15; U.S. Bureau of Reclamation (USBR): Figs. 16, 21, 26, 80-82; B.D. van't Woudt (FAO): Figs. 25, 76, 92, 93; J.D. Zimmerman (A15): Figs. 22, 34-38.

CONTENTS

Preface v

Introduction 1

1. Determining the need for lining 3
    WATER CONSERVATION 3
    PREVENTION OF DAMAGE TO ADJACENT LAND AND DRAINAGE COST REDUCTION 8
    REDUCED DIMENSIONS AND RIGHT-OF-WAY COSTS 10
    REDUCED MAINTENANCE AND OPERATION COSTS 12
    PROTECTION AGAINST EROSION, STRUCTURAL SAFETY AND OTHER BENEFITS 17

2. Determination of seepage losses 18
    FACTORS AFFECTING SEEPAGE 19
    ESTIMATION OF SEEPAGE FROM A PROPOSED CANAL........... 22
        Calculation of seepage losses 22

            Empirical formulae — Analytical solutions — Graphical solutions derived from electrical analogy

        Methods for determining soil hydraulic conductivity 31

            Methods of measuring hydraulic conductivity below the water table — Methods of measuring hydraulic conductivity above the water table — Laboratory methods of measuring permeability

  Conclusions 37
 MEASURING SEEPAGE FROM EXISTING CANALS 38
  Inflow-outflow method 38
   Current meter — Flumes, drops, weirs, orifices
  Ponding method 40
   Examples of ponding tests
  Seepage meter method 44
   Seepage meter with submerged flexible water bag — Falling-head seepage meter — Application considerations
  Special methods 49
   Tracer method — Electrical logging or resistivity measurement method — Piezometric surveys — Remote sensing

3. **Design and construction** 51
 HARD-SURFACE LININGS 51
  General design considerations 51
   Cross section — Subgrade — Subgrade sterilization — Embankments — Drainage — Water velocities — Coefficient of roughness — Prevention of silt deposits in canals
  Cement concrete linings 62
   Design of cement concrete linings — Construction of cement concrete linings
  Pneumatically applied mortar 102
  Grouted fabric mats 103
  Precast concrete slab or block linings 103
  Soil-cement linings 108
   Dry-mix soil-cement — Plastic soil-cement — Methods of testing soil-cement for canal linings — Cost and performance
  Asphaltic concrete linings 116
   Hot-mixed asphaltic concrete — Cold-mixed asphaltic concrete

Prefabricated asphaltic linings	120
Brick linings	121

Examples of brick linings — Construction aspects — Asphaltic or synthetic membranes in brick linings — Clay tiles in small channels

Stone linings	125
EXPOSED MEMBRANES	128
Installation	128
COVERED MEMBRANE LININGS	129
General design considerations	129

Subgrade — Protective cover

Buried sprayed-in-place asphaltic membranes	132
Prefabricated asphaltic membrane linings	134
Plastic and synthetic rubber membrane linings	134

Materials — Installation — Cost — Examples

Fabric-cum-cement plaster lining	144
Bentonite membrane linings	144
EARTH LININGS	145
General design considerations	145
Thick compacted earth linings	147

Suitable soils — Construction considerations

Thin compacted earth linings	155
Compaction of slopes and bed of unlined earth canals and ditches	155
Loosely placed earth linings	156
Soil modification	156
Cost of earth linings	157
SOIL SEALANTS	158
FLUMES, PIPES AND LAY-FLAT TUBING	159
Flumes	159

Pipelines	161
Lay-flat irrigation tubing	164

4. Selection of type of lining ... 166

FACTORS GOVERNING THE SELECTION OF LINING ... 167
- Soil properties ... 167
- Topography ... 173
- Water table ... 173
- Land use and irrigation systems ... 173
- Operation and maintenance ... 174
- Watertightness ... 175
- Durability ... 175
- Availability of construction materials ... 176
- Availability of labour and machinery ... 176
- Cost and financial aspects ... 177

BENEFIT-COST ANALYSIS ... 178

5. Suggestions for further research and development ... 183

Glossary ... 185

Bibliography ... 189

A. General ... 189

B. Seepage losses ... 190

C. Design and construction of lining ... 192

D. Economics of lining ... 198

E. Operation and maintenance of lining ... 198

F. Miscellaneous ... 199

ILLUSTRATIONS

1. Weed-infested irrigation ditch in which seepage and evapotranspiration losses of up to 20 percent per kilometre were measured 6

2. Annual average measured seepage rates over time of study for Trenton sandy loam with twenty parts Redmond bentonite (River Laboratory, Logan, Utah, U.S.A.) 7

3. Annual average measured seepage rates over time of study for buried asphaltic membrane (River Laboratory, Logan, Utah, U.S.A.) 7

4. Clearing aquatic weeds in an unlined canal, Kota, Rajasthan, India 15

5. Weed infestation in a silted and badly maintained concrete canal 15

6. Geometry and symbols for canals under conditions A, B and A′ 27

7. Results of seepage analyses by electric analogy for a trapezoidal canal ($H_w/W_b = 0.75$) 28

8. Results of seepage analyses by electric analogy for a trapezoidal canal ($H_w/W_b = 0.50$) 29

9. Results of seepage analyses by electric analogy for a trapezoidal canal ($H_w/W_b = 0.25$) 29

10. A double-tube apparatus 35

11. Equipment used for the Kopecki-ring method of determining hydraulic conductivity in undisturbed samples 37

12. Measuring flow with a 90° V-notch weir, western Macedonia, Greece 40

xi

13a,b. Ponding tests in distribution ditches with ordinary jute sacks, western Macedonia, Greece 43

14. Seepage meter with submerged plastic bag 45

15. Falling-head seepage meter in a canal 46

16. Bank height for canals and freeboard with hard-surface, buried membrane, and earth linings 53

17-18. Views of breakdown in the Monegros Canal, Spain, 1954   54

19. Typical cross section of hard-surface-lined canal 56

20. Collapse of concrete lining caused by (a) lack of compaction and (b) undersized supporting embankments 57

21. Flap valve installation for a canal underdrain 59

22. Critical velocity curves for the prevention of silt deposits in canals 63

23. Irrigation canal high on a hillside, Horseshoe Feeder Canal, near Loveland, Colo., U.S.A. 64

24. Properly lined and maintained canal, Thailand. Note width of embankments 64

25. Concrete lining in canal on a 10 percent slope, with ladder-type checks for velocity control, Otago, New Zealand ... 65

26. Determination of thickness of hard-surface lining based on canal capacity 66

27. Typical joints for concrete canal lining 69

28. Forming dummy groove contraction joints with a template   70

29. Detail of ASAE standard groove 70

30. Standard trapezoidal canal section 72

31. Capacity chart for concrete-lined canals 73

32. Concrete lining deteriorated by aggressive water during 14 years of service, Mauritius 74

33. Curing of a cement concrete lining, Chambal Project, India   79

34. Parallel boards and screed guides for hand method of lining   80

35. Placing alternate lining sections between screed guides ...   80

36.	Concrete for lining being shovelled against subgrade	81
37.	Concrete being compacted, formed and smoothed with a screed	81
38.	Concreting between set sections, using them as guides	82
39-41.	Steel concrete form for rectangular canal sections	83-85
42.	Detail of template for panel-formed concrete linings	86
43.	Panel form in place and ready for concrete to be poured	86
44.	End spacers in place on alternate sections of panel-formed lining	88
45.	Concrete lining by hand: (*a*) trimming canal section; (*b*) placing 5 cm concrete lining; (*c*) hand screeding; (*d*) final finishing	89
46.	Concrete lining of a canal on the San Francisco River Project, Brazil: (*a*) placement of transverse joints; (*b*) canal subgrade with finished joints; (*c*) placing 12 cm concrete lining; (*d*) completed canal reach	90
47.	Vibrating screed and small float	91
48.	Hand-placed concrete lining of a canal in the Nzoia River Pilot Irrigation Scheme, Kenya	92
49.	Drawing of a type of slipform used to build concrete linings	92
50.	A subgrade-guided slipform	93
51.	Slipform shown in Figure 50 with wheels added to convert it into a berm-guided slipform	94
52-53.	Front and rear views of a bottom-guided slipform for standard concrete lining	94-95
54.	Subgrade-guided slipform from front	96
55.	Subgrade-guided slipform from rear	96
56.	Subgrade-guided slipform equipped with screw hopper	97
57.	Canal slipform paver capable of paving various depths	98
58.	Lining a semicircular canal in Spain	99
59.	Semicircular concrete-lined canal in Spain	100

60. Construction of the main canal, Columbia Basin Project, Washington, U.S.A., showing (in order from foreground) subgrade trimmer, slipform liner, jumbo for cutting dummy groove contraction joints, jumbo for applying membrane-curing compound 100

61. Placing concrete on the slope of Pole Hill Canal, Colorado   101

62. Sliding screed for operating transversely up the slope of a canal 102

63. Precast concrete slab designs 104

64. Lining an irrigation canal with large concrete slabs on polyethylene sheeting, Hunger Steppe, Uzbekistan, U.S.S.R.   105

65. Stages in the lining of a branch canal in India with manually fabricated concrete slabs: (*a*) filling the forms with concrete; (*b*) moulding the concrete slabs; (*c*) curing; (*d*) curing and stockpiling: (*e*) lining site 106-107

66. Form for casting small concrete channel sections 109

67. Hand placing of a soil-cement lining approximately 7.5 cm thick in the Guararé Pilot Scheme, La Villa River Irrigation Project, Panama 111

68. Completed reach of the canal shown in Figure 67 111

69. Two views of an asphaltic concrete slope-lining machine on the North-South Navigation Canal, Fed. Rep. of Germany   119

70. Test core hole in an asphaltic concrete lining 120

71. Double-tile lining, Haveli Canal, India 121

72. Two types of one-tile lining, Sirhind Feeder, India 122

73. Drainage arrangement for brick lining in soils with low permeability 123

74. Combined tile and buried membrane lining 124

75. Stone-lined irrigation canal in Mauritius 126

76. Masonry lining of lava rock mortared together, Hawaii .. 126

77. Stone-lined canal for conveying water from a well, near Dhamar, Yemen Arab Republic 126

78a.	Placing "Katla"-stone slabs in a main canal, Chambal Project, India	127
78b.	"Katla"-stone lining damaged by high velocities and turbulence at a pipe outlet	127
79.	Typical section of buried membrane lining installation	130
80.	Second application of asphaltic membrane with handborne spray bar, Columbia Basin Project, U.S.A.	133
81.	Placing lightweight, buried, glass-fibre reinforced, prefabricated asphaltic canal lining, Altus Project, Oklahoma	134
82.	Unfolding and placing polyethylene (PE) plastic lining from a 30 m roll	137
83.	A dragline carefully dumping earth on a vinyl liner, Nebraska	138
84.	Installing PVC-membrane lining on a main feeder canal, Kirkuk Irrigation Project, Iraq: (*a*) section of membrane in place; (*b*) joining the membrane; (*c*) covering the membrane with a mixture of silty sand and gravel; (*d*) completed section covered with layer of riprap	139-140
85.	Damage to a covered PVC-membrane lining caused by flood water overtopping the levee, Kirkuk Irrigation Project, Iraq	141
86.	(*a*) Section of water course lined with a buried polythene sheet. (*b*) Section of masonry profile wall	142
87.	Section of water course lined with polythene sheet overlaid by brick masonry	143
88.	Typical section of a compacted-earth-lined canal	146
89.	Compaction by tractor-drawn sheepfoot roller on a 4:1 slope	153
90.	Excavation of an irrigation ditch in compacted earth, Chambal Project, India	154
91.	Compacted earth canal after nine months in operation, Chambal Project, India	154
92.	Wooden flume for water distribution to furrows on sugarcane land, Hawaii	159
93.	Precast concrete flume with siphon under a farm road, Japan	160

94.	Concrete flume, Spain	160
95.	Flume of sandstone slabs (Katla) and bricks, Rajasthan, India	161
96.	Drawing of a buried-pipe delivery system	162
97.	Gated tubing to irrigate a tomato crop	164
98.	Some methods of irrigating crops with lay-flat tubing	165
99.	Comparative cross sections of an unlined and a concrete-lined canal	180

TABLES

1. Estimated water losses in unlined conveyance systems .. 4–5
2. Seepage losses for unlined canals and some typical linings .... *Foldout after* 8
3. Maximum non-erosive velocities 10
4. Maximum non-erosive velocities in earth canals 11
5. Relative capacities of concrete-lined and unlined canals 11
6. Cross-sectional data for canals on some federal irrigation projects in the United States 13–14
7. Increase of seepage with increase of water depth in a canal 21
8. Magnitude of specific hydraulic conductivity 31
9. Manning's coefficient of roughness (n) for unlined and lined canals 60–61
10. Dimensions of standard trapezoidal canal sections 71
11. Approximate cement concrete mixes 77
12. ASAE specifications for slipform concrete 78
13. Average cement requirements of various materials 114
14. Examples of soil-cement linings 115–116
15. Suggested mix compositions for dense-graded asphaltic concrete linings 117
16. Characteristics of commonly used polymer membranes . 136

xvii

17. Important physical properties of soils and their uses for canal linings based on the Unified Soil Classification System 149–150

18. Average properties of soils 151

19. Irrigation canal linings and their main features 168–172

INTRODUCTION

Conveyance and distribution of water are an integral part of an irrigation project. Water obtained from natural streams and reservoirs must often be conveyed in canals 100 km or more in length. On some projects days are required to convey water from points of diversion to points of use; on many, 100 km of canal networks are required for each hectare of irrigated land.

The efficiency of the conveyance and distribution system — that is, the transport of water at minimum cost and with minimum water loss — therefore essentially affects the total economy of an irrigation project. The most important decisions to be made in this respect are whether or not and how to line a canal. Regardless of project size, such decisions must be reached by the same kind of reasoning; the criteria used may or may not be identical.

Determination of the need to line should be based on an analysis of benefits such as water conservation, reduced waterlogging or lower drainage requirement, reduced excavation and right-of-way cost, lower operation and maintenance costs, and structural safety.

Materials used for canal lining are of almost infinite variety, but they can be generally categorized as hard-surface, membrane and earth linings. No particular type of lining is lowest in cost or most satisfactory for use in all locations; almost every type of lining has its specific merits. Selection should be based on a careful analysis of local conditions such as availability and cost of labour, mechanical equipment and construction materials; transport facilities; anticipated irrigation method and canal operation; traditional lining techniques; and maintenance requirements. The availability of labour, skilled and unskilled, is perhaps the most important factor causing differences in lining practices. For instance, in some cases it would be unthinkable to use highly mechanized methods in an area where many people are in need of employment. Such circumstances may make one type of lining unfeasible, but at the same time may bring a superior type within economic reach.

1. DETERMINING THE NEED FOR LINING

If the value of the beneficial objectives of lining as they apply to any given case can be reliably estimated, it is possible to determine whether or not lining is needed. In some cases the justification for a certain type of lining may be so obvious that no thorough benefit-cost analysis is necessary. In most cases, however, a proper evaluation of all benefits and their correlation to the initial and current cost of lining is necessary.

In this chapter the possible tangible and intangible benefits of canal lining, as well as means for their evaluation, are discussed. The subject of seepage losses, because of its extent and importance, is treated separately in Chapter 2. Criteria for the selection of type of lining best suited to given conditions are discussed in Chapter 4.

Water conservation

Conservation of water supplies is becoming increasingly important all over the world as the demand for this vital commodity continues to rise rapidly and new sources of supply become scarcer. The time is soon approaching when the only additional natural water supplies will be those available through the salvage of those now being lost. One of the most important ways in which full use of natural water supplies for agriculture can be achieved is through a reduction in the amount of water lost by seepage during transportation to farmers' fields and through weed control.

Water losses in unlined conveyance systems are usually high. These are illustrated in Table 1 which shows estimates from various sources.

The percentage of seepage losses in small canals and farm ditches is normally greater than in large conveyance canals: in those carrying from 30 to 140 litres per second (1 to 5 cusecs), water losses from seepage and water consumption by weeds (Figure 1) can be as high as 20 percent per 1.6 km (= 1 mile) (A2). The author measured an average of 17.5 percent/km in small irrigation channels of 20-100 litres per second capacity in Greece (western Macedonia).

TABLE 1. — ESTIMATED WATER LOSSES IN UNLINED CONVEYANCE SYSTEMS

Reference	Country (project)	Water losses as percent of total water diverted	Remarks
U.S. Bureau of Reclamation (A12)[1]	46 irrigation projects in the U.S.A.	3-86 (average 40)	Records from 46 irrigation projects, including seepage water taken up by uncontrolled vegetation in canals and evaporation losses of canals
Khangar (F6)	West Pakistan	18-44	Seepage losses only
Maasland, M.	West Pakistan: Indus River Basin	35	Mean figure of total conveyance losses
Kennedy (B40)	West Pakistan: Bari Doab Canal	20 6 21 ――― 47	Canals and branches Distributaries Water courses (ditches) Total losses
Barona, F. (A3)	Mexico	26 35-50	Less pervious soils More pervious soils
Doneen, L.D. (B2)	Turkey: Konya Cumra Plain Menemen Plain	40 30	
Lauritzen, C.W. (C16)	Egypt: Nile Delta area New canals in desert areas	8-10 50	Low because of silting effect of Nile water
Sharov, I.A. (E7)	U.S.S.R.	20-35	Mains and distributaries
Sain, K. (A3)	India: Ganges Canal	15 7 22 ――― 44	Main canals and branches Distributaries Water courses Total seepage losses

[1] References to bibliography are thus indicated throughout table.

TABLE 1. — ESTIMATED WATER LOSSES IN UNLINED CONVEYANCES SYSTEMS (*concluded*)

Reference	Country (project)	Water losses as percent of total water diverted	Remarks
EPTA Report No. 1519 1962 (B40)	Pakistan: Kushita unit of the Ganges-Kobadak Irrigation Scheme	Maximum 40 5.7 7.3 12.0 ——— 25.0	Total seepage losses Main canals Secondary canals Tertiary canals Total seepage losses
Hekket, H. 1969	Iran: Garmsar Irrigation Project	40	Main and secondary canals
Ministry of Public Works, Chile	Huasco Valley Project	54 (ca. 2.2/km)	Canal 25 km in length having about 1 m^3/sec discharge capacity
ICID (A5)	U.S.S.R.: Kara Kum Canal, 400-km reach; 28 to 6 m; sandy soils	43	Average loss during first year of operation; losses decreased in subsequent years as water table rose
ICID (A5)	Algeria: El Arjiane	40	Average losses in channels dug in sandy soil
Irrigation and Power Department, Punjab, Pakistan (D10)	Punjab Province	11	Average losses in 44 000 water courses equal to 7 000 million m^3 per year

Lining a canal will not completely eliminate losses; therefore it is necessary to measure systematically present losses or estimate the losses which might reasonably be saved by lining before a proper decision can be made. Some authors state that roughly 60 to 80 percent of the water

FIGURE 1. Weed-infested irrigation ditch with measured evapotranspiration losses of up to 20 percent per kilometre.

lost in unlined canals can be saved by hard-surface lining (A3, C2). As a rule of thumb, a canal properly lined to reduce seepage should not lose more than about 30 $l/m^2/$ 24 h (0.1 $ft^3/ft^2/24$ h). This is roughly a loss of 0.6 percent/km of the conveyed water in a canal carrying 140 l/sec (5 cusecs).

Houk (A9) states: "Ordinarily, expenditures involved in reducing canal seepage losses to values as low as 5 percent of total conveyed water are economically justified only in regions where water supplies are insufficient and crop products are unusually valuable."

Under certain conditions where water lost through canal seepage may be recovered and used within the project area, the total cost of preventing such seepage should not be charged to the benefit side of the economic comparison.

Table 2 (following page 8) shows measured or calculated seepage losses in various unlined and lined canals, demonstrating the scope of possible seepage reduction with different types of lining. The table may also serve as a guide in determining the need for lining in cases where specific figures on seepage losses cannot be obtained by tests, calculations or comparisons.

Figures 2 and 3 show the annual measured increases in seepage losses for bentonitic membrane and buried asphaltic membrane linings during tests conducted by the River Laboratory, Logan, Utah, U.S.A.

Emergent aquatic weeds and phreatophytes transpire large quantities of water from canals which may also be saved by lining. Measurements indicate that some water-loving natural vegetation uses from 50 to 100 percent more water than most field crop plants. A survey conducted in 1948 in the U.S.A. showed that nearly 185 million m^3 (150 000 acre-feet) of water were lost each year in the Bureau of Reclamation's 22 600 km (14 075 mi) of canals because of weeds (E4). Detailed procedures for estimating the consumptive use by these plants are given in "Consumptive use and water waste by phreatophytes," by H.F. Blaney (ASCE

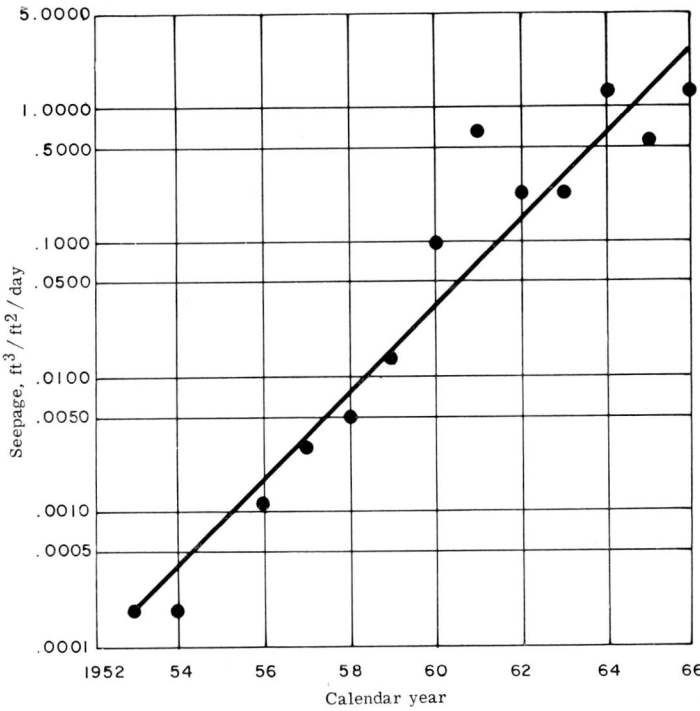

FIGURE 2. Annual measured seepage rate over time of study for Trenton sandy loam with twenty parts Redmond bentonite (River Laboratory, Logan, Utah, U.S.A.)

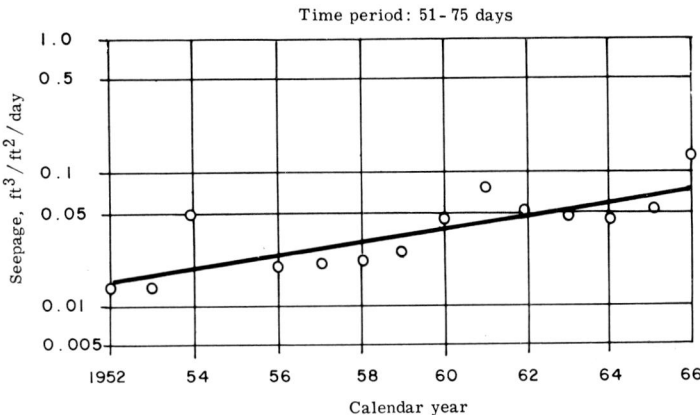

FIGURE 3. Annual measured seepage rate over time of study for buried asphaltic membrane (River Laboratory, Logan, Utah, U.S.A.).

Proceedings No. 2929, September 1961). The benefit of saving water consumed by weeds may contribute to justifying the use of hard-surface linings which have proved weed-resistant.

Conservation of water has little significance unless the water saved can be put to beneficial use or its loss creates problems with regard to human needs. When the water saved can be controlled and delivered to productive lands which would otherwise produce at less than full capacity, its value can be estimated to assess the benefit of lining. Even so, it is sometimes difficult to estimate the value of additional irrigated lands in relation to the overall value of a project, community or national. The minimum measurable return from additional land receiving water or additional yield on existing land is the net income of the individual farmers. The maximum, although indeterminate, exceeds the gross income since additional people derive benefits from the production and consumption of products from the land (A8).

Prevention of damage to adjacent land and drainage cost reduction

The extent of canal seepage and its influence on land drainage problems are difficult to measure. Seepage often disappears into a pervious underground stratum and reappears in a lower area some distance from the canal. It is difficult to accurately determine prior to project construction where groundwater will be a problem. In areas subject to high water tables it is sometimes difficult to isolate causes. Lining of main canals may only partially alleviate problems unless sub-laterals and farm ditches are also lined and farmers apply only that amount of water which can be used and removed without overloading the drainage system. Therefore, careful analysis must be made, taking into consideration the natural drainage of the area, the expected farm irrigation efficiencies, the probable salt problems, and loss of land and its economic and human implications if high water tables occur owing to excessive canal losses. It must also be borne in mind that lowering the water table will increase seepage from unlined canals, as has been confirmed in the field as well as in the laboratory (A6).

Some examples may illustrate the magnitude of lining benefits with regard to adjacent land conditions.

On one project in the western United States some 8 500 hectares of cultivated land had become so waterlogged by seepage from canals and laterals that they had to be abandoned. Numerous open drains had been constructed, but these were not sufficiently effective since much good farm land continued to be waterlogged and many farmsteads, including

NOTES TO TABLE 2

1. Average conditions for canals not affected by the rise of groundwater owing to seepage. The higher values are for comparatively new canals. Seepage loss usually decreases noticeably with age, particularly if the water carries sediments (A7).

2. Data compiled by the U.S. Bureau of Reclamation on actual seepage losses in canals excavated through certain types of soil. They represent average losses from observations on eight different projects (A9).

3. Measured by ponding test in an unlined 17 000 ft (5 182 m) reach of the Right Main Canal, Chambal Commanded Area, India (B26). Water table: 3 ft (ca. 1 m) below canal bottom; mean wetted perimeter: 167 ft (51 m). Canal had been in service for eight years.

4. Average of six ponding and six seepage-meter tests on the Interstate Lateral 24A, North Platte Project, Wyoming, U.S.A. (A1).

5. East Contra Costa Canal, basin 2 + 3, U.S.A. (A9).

6. E-W Farm Lateral, Fort Collins, Colorado, U.S.A. (A9).

7. West Canal, Rio Grande Project, New Mexico-Texas, U.S.A. Average figure from three test reaches (A1).

8. Bhakra Main Canal, India (C2). Bed width: 52-63 ft (16-19 m); water depth: ca. 20 ft (6 m).

9. Mean seepage rate on main canal: approximately 110 m^3/sec; capacity: 4-5 m deep. Mexico (A3).

10. Lateral R-4-S, Shoshone Project, Wyoming, U.S.A. (A1). Design capacity: approximately 85 ft^3/sec (2.4 m^3/sec); water depth: 2.1-2.2 ft (0.64-0.67 m).

11. Two reaches of the Bow River–F-5 Lateral, Canada, excavated into loam on sand and gravel. No information on canal size (B30).

12. Conchas Canal, Tucumcari Project, New Mexico, U.S.A. (A1). Canal capacity: 700 ft^3/sec (19.8 m^3/sec); water depth as tested: 7 ft (ca. 2 m).

13. Supply canal, Abda Donkkala, Morocco. Canal capacity: 16 m^3/sec. Losses before lining refer to sandy regions "where the land was very porous" (A3).

the basements of farmhouses, gradually became flooded. Lining placed in canals and in many principal laterals have now reduced the seepage to the point that cropping is again possible in a number of fields. Some open drains have been filled in and the area occupied by them has been returned to cultivation (A1).

In an irrigation district near San Juan, Texas, 56 km of a canal were lined with concrete before 1938. In addition to saving valuable water, 1 200 hectares of high-priced orchard land which had become waterlogged by seepage were reclaimed (A2).

On the Milk River Project in Montana the placement of a lining eliminated a serious problem. Here seepage was so extensive that the banks of the nearby river were caving and sloughing so badly as to endanger an adjacent railway line (A1).

India's Committee on Natural Resources reports that waterlogging made its first appearance in the Punjab in the area irrigated by the Western Jamuna Canal around 1850. The effects were a considerable decrease in crop yields and deterioration in the general health of the people caused by malaria and other diseases. Investigations showed that, as a result of the waterlogging and poor alignment of the canal, the water table had risen appreciably. Between 1870 and 1880 the canal was re-aligned and the natural drainage improved by the provision of adequate waterways, which brought about a marked lowering of the water table (F2).

Reduced dimensions and right-of-way costs

In canals lined with exposed hard-surface materials such as cement concrete, asphalt concrete, stone and brick masonry and certain other types of lining, greater velocities are permissible than are normally possible in earth canals. The maximum non-erosive velocities for different soils are shown in Tables 3 and 4. They range roughly from 0.3 to 1.8 m/sec

TABLE 3. — MAXIMUM NON-EROSIVE VELOCITIES (A15)

Fine sand under quicksand condition	0.20—0.30 m/sec	(0.75—1.00 ft/sec)
Sandy soil	0.30—0.75 m/sec	(1.00—2.50 ft/sec)
Sandy loam	0.75—0.90 m/sec	(2.50—3.00 ft/sec)
Loam to clay loam	0.85—1.10 m/sec	(2.75—3.75 ft/sec)
Stiff clay	1.10—1.50 m/sec	(4.00—5.00 ft/sec)

TABLE 4. — MAXIMUM NON-EROSIVE VELOCITIES (M/SEC) IN EARTH CANALS (E7)

Nature of canal bed	Having a discharge in m³/sec of:								
	0.5	1.0	2.0	3.0	4.0	10.0	15.0	20.0	30.0
Silt, fine sand, light sandy loam	0.37	0.39	0.41	0.43	0.45	0.47	0.49	0.50	0.52
Medium sandy ground	0.46	0.49	0.52	0.54	0.56	0.59	0.61	0.63	0.65
Light loam	0.53	0.56	0.59	0.61	0.64	0.68	0.70	0.72	0.74
Medium loam, medium loess, coarse sand	0.59	0.63	0.67	0.69	0.72	0.75	0.79	0.81	0.84
Heavy loam, light clay, close grain loess, very coarse sand	0.67	0.71	0.75	0.78	0.81	0.86	0.89	0.90	0.94
Fine shingle or gravel	0.73	0.77	0.82	0.84	0.88	0.93	0.96	0.98	1.02
Thick medium clay, medium gravel	0.82	0.87	0.92	0.95	0.99	1.05	1.09	1.11	1.16
Heavy clay (tertiary), coarse shingle or gravel	1.26	1.34	1.42	1.47	1.53	1.62	1.68	1.72	1.79

(1 to 6 ft/sec), while the flow velocities for concrete and brick linings range from 1.5 to 2.5 m/sec (5 to 8 ft/sec) (A3). Although clay stands velocities up to 1.5 m/sec in a newly excavated canal, in time the flow resistance may drop considerably owing to alternate wetting and drying effects and other structural changes. (See also Chapter 3.)

Table 5 shows the relative quantities of water which can be transported through a concrete-lined canal and an unlined canal of the same size. To supply a given discharge the surface area of the concrete lining can

TABLE 5. — RELATIVE CAPACITIES OF CONCRETE-LINED AND UNLINED CANALS (A2)

Bottom width	Flow depth	Quantity (m³/sec)	
		Concrete-lined	Unlined
0.30	0.45	0.40	0.23
0.90	0.60	1.27	0.71
1.20	0.75	2.40	1.33
1.50	0.90	4.00	2.24

be reduced as the friction loss is less and the permissible velocity is greater.

When erosion-resistant linings (such as concrete) are used, the gradient of the bed can be increased and the side slope made steeper. The narrower width reduces the land area required for the canal, and the right-of-way cost savings may be substantial.

Considering that a lined canal system for a given supply area need not carry the water which is lost in an unlined system, the dimensions of any lined canal system, including structures, will be smaller than those of an unlined system.

Hard-surface lining also provides flexibility in canal location. Canals can follow abrupt contours since the erosion hazard associated with increased water velocities on the outer edge of the canal is controlled. Lined canals can be used on steeper gradients when advantageous, which may eliminate the drop structure necessary for erosion control in unlined canals of the same slope.

Greater velocities in canals reduce maintenance costs when silting is a problem. The silt, by remaining in suspension, does not fill the canal but rather settles out on the land being irrigated. Efforts should be made to remove the silt before it enters the distribution system.

Greater conveyance velocities resulting in increased discharges offer the advantage of shorter irrigation time.

Table 6 shows cross-sectional data for some concrete-lined and unlined canals on federal irrigation projects in the western United States, to illustrate the different characteristics of lined and unlined canals.

Reduced maintenance and operation costs

The type of lining must be considered when evaluating the benefits of canal lining with reference to maintenance costs. For example, the economic benefits of using hard-surface lining are reduced costs of weed control, less danger from burrowing animals, less silt removal, and other conditions which a rigid, high-quality lining will provide. But it should be remembered that economic studies for an earth or buried membrane lining cannot include many of these factors and must rely on the value of other benefits for justification.

One of the largest recurring maintenance costs in many canal systems is weed control and the removal of weeds and water-loving plants from the canal section (Figure 4). High-quality hard-surface linings such as concrete, shotcrete, tile, stone and to a lesser extent asphaltic concrete, being relatively impenetrable by weeds and water-loving plants,

TABLE 6. — CROSS-SECTIONAL DATA FOR CANALS ON SOME FEDERAL IRRIGATION PROJECTS IN THE UNITED STATES

Canal	Project	Side slopes	Depth of flow	Bottom width	Ratio: width to depth	Mean velocity	Discharge	Freeboard
Unlined earth sections			.... ft			ft/sec	ft³/sec	ft
Lateral	Altus	1½:1	1.66	4	2.41	1.86	20	1.3
C Line East	Boise	1½:1	4.00	8	2.00	1.89	106	3.5
C Line East	Boise	1½:1	7.14	14	1.96	2.25	397	3.4
Altus	Altus	1½:1	6.20	20	3.23	2.48	450	3.5
Conchas	Tucumcari	1½:1	8.65	24	2.77	2.19	700	4.3
Kittitas Main	Yakima	1½:1	11.35	30	2.64	2.47	1 320	5.0
Ridge	Yakima	1½:1	9.57	40	4.18	2.50	1 300	4.0
Main (Gravity Extension)	Minidoka	1½:1	5.60	60	10.71	3.00	1 149	3.0
Coachella	All American	2:1	10.33	60	5.81	3.00	2 500	6.0
Gravity Main	Gila	2:1	13.54	100	7.38	3.49	6 000	6.0
All American	All American	2:1	16.59	130	7.84	3.75	10 155	6.0
All American	All American	1¾:1	20.61	160	7.76	3.75	15 155	6.0
Concrete-lined earth sections								
Contra Costa	Central Valley	1¼:1	1.84	3	1.63	2.33	26	1.16
South Branch	Yakima	1¼:1	3.80	5	1.32	5.94	220	1.00
Ridge	Yakima	1¼:1	7.46	7	0.94	4.93	600	1.04
Heart Mountain	Shoshone	1¼:1	7.51	8	1.06	7.00	914	1.49
Kittitas Main	Yakima	1¼:1	8.99	11	1.22	6.63	1 320	1.25
Black Canyon	Boise	1¼:1	9.39	12	1.28	4.86	1 089	1.61
Ridge	Yakima	1¼:1	11.20	14	1.25	7.02	2 200	1.80
All American	All American	1⅔:1	10.64	22	2.07	6.62	2 800	2.25
Delta-Mendota	Central Valley	1½:1	16.56*	48	2.90	—	4 600	—

* To top of bank.

TABLE 6. — CROSS-SECTIONAL DATA FOR CANALS ON SOME FEDERAL IRRIGATION PROJECTS IN THE UNITED STATES (*concluded*)

Canal	Project	Side slopes	Depth of flow	Bottom width	Ratio: width to depth	Mean velocity	Discharge	Freeboard
Unlined rock sections			.... ft			ft/sec	ft³/sec	ft
North Unit Main	Deschutes	¼:1	5.54	20	3.61	4.63	550	2.5
Gravity Main	Gila	¾:1	21.57	33	1.53	1.79	1 900	Deep cut
Yuma	Yuma	½:1	8.46	60	7.10	3.68	2 000	Deep cut
All American	All American	¾:1	20.13	69	3.42	6.00	10 155	Deep cut
All American	All American	¾:1	22.75	94	4.13	6.00	15 155	Deep cut
Concrete-lined rock sections								
Kittitas Main	Yakima	½:1	9.00	14	1.56	7.65	1 275	1.25
Main (Gravity Extension)	Minidoka *	¼:1	13.80	20	1.45	8.20	2 695	1.50
Gravity Main	Gila	¾:1	21.07	32	1.52	5.96	6 000	—

* To top of lining.

greatly reduce the cost of weed control and removal from the canals. Any of the hard-surface linings will allow water velocities high enough to reduce substantially the deposit of silt, thus lowering maintenance cost. However, the deposit of silt and other material may be a problem during off-seasons. Figure 5 shows a canal which has been silted up by soil washed from the side slopes during heavy rains.

Buried membrane linings or compacted earth linings utilizing a substantial gravel blanket prevent or substantially reduce erosion, which might be a problem in an unlined canal. The gravel blanket also discourages burrowing animals and provides less favourable conditions for weed growth near the water's edge. Benefits which it would seem reasonable to attain with these linings should be considered in the economic analysis of their feasibility.

Maintenance costs are greatly dependent upon climate, period of operation, availability of labour and machinery, type of lining, canal charac-

FIGURE 4. Clearing aquatic weeds in an unlined canal, Kota, Rajasthan, India.

FIGURE 5. Weed infestation in a silted and badly maintained concrete canal.

teristics and so on. Therefore, comparison of maintenance expenses between projects is not generally practicable. Estimation of maintenance costs can be obtained from data for existing lined and unlined canals operating under similar climatic, geographic and agricultural conditions.

The following data on maintenance costs may serve as a guide for estimating lining benefits.

In the United States studies of operation and maintenance costs for 2 050 km (about 1 300 miles) of lined and unlined canals in various parts of the country, based on reports covering a two-year period, showed a 75 percent reduction when hard-surface linings were used (A4).

In another study, in which conditions of the compared systems were quite similar, the annual operation and maintenance costs of a pressurized concrete pipe system were about half those of an unlined ditch system.

Joseph (D6), writing on India, says that, on the average, two man-days are necessary for maintaining about 40 m of unlined field canal in a year, including cleaning weed growth, repairing breaches, strengthening bunds, clearing silt and making fresh outlets. This refers to an 8-hectare farm with a private tubewell of 14 l/sec capacity for irrigation and a distribution system about 750 m in length. With lined canals and control structures there should be a saving of about 50 percent in labour.

Wineland and Lucas (B4) assumed $2 per metre of canal per year as the difference in maintenance and operation cost between a thick compacted-earth-lined and a concrete-lined canal of equal capacity, ranging between 37 and 23 m^3/sec. The cost included lining repairs, canal cleaning, maintenance of embankments and structures, and weed control.

Some records in the U.S.S.R. indicate that for concrete-lined canals the saving in maintenance costs may be as much as two thirds the cost of maintaining earth canals of similar capacity (A13).

A survey conducted in 1957 in the United States showed that the average cost per mile (1.6 km) for treating aquatic weeds ranged from $11.08 to $121.42, averaging $42.36 in seven western irrigation systems. The control methods used included drying, handcutting and cleaning, chaining, draglining and treating with chemicals. The cost of ditch-bank weed control averaged $53.00 per hectare for all regions, ranging from $15.50 to $137.00. Methods of control included handcutting, mowing, burning, and spraying with many different kinds of equipment (D9).

In the lower plain of the Segura, Spain, a study was made to obtain average figures for the savings which could be made in annual canal maintenance by comparing the corresponding cost of lined and unlined canals of different size and location. It was concluded that lining resulted in savings equivalent to 0.34 man-days per running metre of canal per year (F7).

Protection against erosion, structural safety and other benefits

The stability of the slope and bottom of unlined canals and ditches is a major problem in areas of sandy and silty soils, particularly where the irrigation practices require an intermittent regime in the canal. Protection against erosion under such conditions may be all that is required from lining. Lining will also reduce the danger of canal breaks resulting from erosion, burrowing animals or slippage. This is particularly important when a canal is to be located on a fill or a steep side slope.

Another intangible but, under certain circumstances, important benefit of any hard-surface or membrane lining is that the lining prevents the water from absorbing salts which may be harmful to crops from the soil.

Substantial savings may also be effected by reduced pumping costs due to more efficient water use. For the Ganges-Kobadak Irrigation Scheme, India, it was estimated that savings in pumping costs alone would justify canal lining (D4).

2. DETERMINATION OF SEEPAGE LOSSES

Although erosion resistance, reduced maintenance, reduced right-of-way cost and safety may justify lining in special cases, seepage prevention is normally the governing factor in lining considerations. Here seepage refers to the process of water movement from a canal into and through the bed and wall material.

Knowledge of seepage rates is required for economic evaluation and design. Seepage rates are obtainable either by estimation or by direct measurement. Estimation is based on knowledge of the relevant hydraulic properties of the soil and of the boundary conditions, such as depth to groundwater, canal cross section and water depth. Because of the many variables involved, no general law for calculating the rate of seepage has been developed.

Methods of evaluating seepage from existing canals are:

(a) discharge measurements by the inflow-outflow technique;

(b) measuring the rate of water loss from a ponded canal;

(c) measuring the rate of water movement into the bottom or bank of the canal with a seepage meter;

(d) special methods essentially limited to qualitative indication of seepage (for example, its distribution along a canal).

Of the many terms used to express amount of seepage, the following are the most generally useful:

— volume per unit area of wetted perimeter per 24 hours ($m^3/m^2/24$ h or $ft^3/ft^2/24$ h);

— volume per unit length of canal per 24 hours ($m^3/m/24$ h or $ft^3/ft/24$ h);

— cubic feet per second per million square feet of wetted perimeter;[1]
— percentage of total flow per kilometre (mile) of canal.

When comparing figures on seepage losses in lined and unlined canals, attention should be paid to the following: for equal unit loss the total volume lost per unit length of canal is greater for an unlined than for a hard-surface lined canal as the wetted perimeter is less for the lined system. For example, the wetted perimeter of a concrete-lined canal is about 30 percent less than that of an unlined canal.

Factors affecting seepage

The main factors known to have a definite effect on seepage rate can be grouped as follows:
1. Characteristics of the soil of the region through which the canal runs.
2. Depth of water in the canal, wetted perimeter of the canal and depth to groundwater.
3. Amount of sediment in the water, velocity in the canal and length of time the canal has been in operation.

Since the effects of these factors on seepage rate are closely interrelated, it is practically impossible to separate them. For the interpretation of test results, however, it is important to know in some detail how the main factors cause seepage.

GROUP 1

The most important soil characteristic is the permeability of the native canal material. Permeability is influenced both by pore size and by percentage of pore space (porosity). Soils consisting of a mixture of gravel and clay are almost completely impervious, while coarse gravel may transmit water many times faster; thus a wide range of seepage losses is possible (see also Table 2 between pages 8 and 9).

---

[1] Equivalents: 1 m^3/m^2/24 h = 3.2816 ft^3/ft^2/24 h. — 1 ft^3/ft^2/24 h = 0.3047 m^3/m^2/24 h. — 1 ft^3/sec/million ft^2 = 0.0864 ft^3/ft^2/24 h.

The survey of the soil profile along the proposed canal is perhaps the most important single technical step in a preconstruction seepage investigation. This survey is made to determine the location, extent and physical characteristics of the various underlying soil layers. The sequence of permeable and impermeable strata in the area and the capability of these strata to transmit water largely determine the amount of water lost by seepage.

GROUP 2

Bouwer (B10) has found the following relationships between seepage and water depth in the canal, depth to groundwater and wetted perimeter of canal:

1. Seepage losses increase with the increase of water depth in the canal.

2. Seepage losses increase as the difference between water level in the canal and water table increases; when the difference is five times or more the surface width of the canal, seepage losses reach the upper limit or the infinity condition.

3. The distribution of seepage losses across the canal bed and sides depends upon the position of the water table or impermeable layer. When the water table is shallow, the seepage from the sides is greater than that from the bed, and the reverse is true with a deep water table. In all cases the maximum seepage losses occur at the toe of the slope — that is, at the junction of the bed and sides of the canal.

4. The significant depth within which the nature of the soil affects seepage losses has been found to be five times the bed width of the canal. Laterally, at a distance ten times the bed width of the canal, the effect of seepage losses on the original water table is insignificant.

Item 1 is illustrated by Table 7, derived from ponding tests in a reach of the Courtland Canal, Missouri River Basin Project, Nebraska. Tests were made (*a*) on the unlined canal in loessial soil, and (*b*) after having compacted the canal bottom only, to a depth of two feet.

In accordance with item 3, it is advisable to construct wide, shallow canals in areas with a high groundwater table and narrow, deep canals in areas with a low water table.

TABLE 7. — INCREASE OF SEEPAGE WITH INCREASE OF WATER DEPTH IN A CANAL

Depth of water (ft)	Seepage rate ($ft^3/ft^2/24$ h)		Approximate reduction in seepage rate (percent)
	Before compaction (Autumn 1962)	After compaction (Spring 1963)	
8.5	1.21	1.06	12
7.5	1.06	0.92	13
6.5	1.90	0.76	16
5.5	0.73	0.57	22
4.5	0.55	0.37	33

GROUP 3

Material suspended in canal water is carried by seepage water into the pores in the soil in which the canal is constructed. If the water contains considerable amounts of suspended material, the seepage rate may be reduced in a relatively short time. Even small amounts of sediment will have sealing effects over a period of time. If the velocity is reduced, the sediment-carrying capacity of the water decreases, resulting in settlement of part of the suspended material. This forms a thin, slowly permeable layer along the wetted perimeter of the canal which decreases the seepage.

For the Interstate Canal, Nebraska, it was estimated that a slight mud content of a maximum of 1 000 ppm would reduce the over-all seepage losses by 20 percent. In the unlined canals in the Kushita Unit of the Ganges-Kobadak Irrigation Scheme, the reduction of losses due to the deposition of suspended sediment was estimated to be 20 percent.

In the 17 km long canal of Donzère-Mondragon, France, seepage losses immediately after construction amounted to 16 m^3/sec, but were reduced to 3 m^3/sec within five years of operation by the natural sealing effect of the silt-laden water of the Rhône (A15).

If the bed material contains expansive clays, seepage will be reduced considerably when these soils become saturated by the seeping water (see *Soil sealants*).

In seasonally used unlined canals the seepage rate will be high at the beginning of the season and gradually decrease toward the end. In most lined canals seepage increases with time for a variety of reasons, depending on the lining material.

Estimation of seepage from a proposed canal

Estimation of seepage may be required for the economic evaluation of the benefits from lining a proposed canal as well as for the design of the irrigation system, including structures. The simplest method of prediction would be to adopt known seepage losses from canals of similar size and shape embedded in soils of similar permeability with a similar groundwater table. Where such comparative data are available, they may be used to estimate the magnitude of seepage losses (see also Chapter 1).

A method of qualitatively predicting seepage losses consists of estimating or measuring the hydraulic conductivity (permeability), K, of the soil in which the canal is to be excavated. The values of K obtained will indicate the distribution of seepage losses along the proposed canal, but not their magnitude.

A quantitative prediction of seepage losses can be obtained by calculation. Calculation methods for various conditions are briefly reviewed in the following paragraph. (Methods of measuring the hydraulic conductivity of soil are discussed on pages 31–37.)

CALCULATION OF SEEPAGE LOSSES

A variety of methods have been developed for the calculation of seepage from irrigation canals. The most useful may be grouped thus:

(*a*) empirically developed formulae;

(*b*) solutions arrived at by analytical methods;

(*c*) solutions derived from electrical analogy.

The use of empirical formulae can only produce rough estimates, whereas analytical methods give highly accurate results when applied to conditions for which they have been developed. However, analytical methods are generally quite elaborate. Bouwer (B10) has studied seepage problems on electrical resistance networks and has from this developed graphical solutions covering a wide range of possible seepage conditions. It is believed that these solutions today represent the most accurate and convenient means of determining seepage values from known hydraulic conductivity of the subsoils, geometry of the canal and position of the groundwater table. The main criticism of analogue methods is the assumption of homogeneous natural conditions, which seldom occur. Investigations of seepage under non-steady-state flow conditions and other seepage problems are being carried out with digital computers and electrical analogies.

Empirical formulae

(*i*) Davis and Wilson (see Dhillon, B18) have suggested the following relationship for the estimation of seepage losses in lined canals:

$$S_L = 0.45 \times C \times \frac{P_w \times L}{4 \times 10^6 + 3\,650\,\sqrt{v}} \times H_w^{1/3}$$

where

S_L = seepage losses (m³ per length of canal per day);
L = length of canal (m);
P_w = wetted perimeter (m);
H_w = water depth in the canal (m);
v = velocity of flow in the canal (m/sec);
C = constant value depending on lining.

Type of lining and thickness	Value of C
Concrete (10 cm)	1
Mass clay (15 cm)	4
Light asphalt	5
Clay (7.6 cm)	8
Asphalt or cement mortar	10

(*ii*) After obtaining the results of surveys on eight different canal systems, the U.S. Bureau of Reclamation proposed the following relationship, known as the Moritz formula (referred to by Dhillon, B18):

$$S = 0.2 \times C \times \sqrt{Q/V}$$

where

S = seepage losses (cusecs per mile length of canal);
Q = discharge (cusecs);
V = velocity (ft/sec);
C = constant value depending on soil type.

Soil type	Value of C
Cemented gravel and hardpan with sandy loam	0.34
Clay and clayey loam	0.41
Sandy loam	0.66
Volcanic ash	0.68
Sand or volcanic ash or clay	1.20
Sandy soil with rock	1.68
Sandy and gravelly soil	2.20

(*iii*) In India the following formula has been used (A5):

$$S = c \times a \times d$$

where

S = total loss (ft^3/sec);

a = area of wetted perimeter (million ft^2);

d = water depth in the canal (ft);

c = a constant.

Observations made on some of the important canals in the Punjab showed that c ranged from 1.1 to 1.8.

(*iv*) For estimating conveyance losses of canals, Egypt's Irrigation Department uses the empirical formula of Molesworth and Yennidumia (B21):

$$S = c \times L \times P \times \sqrt{R}$$

where

S = conveyance losses [m^3/sec per length (L) of canal];

L = length of canal (km);

P = wetted perimeter (m);

R = hydraulic mean depth (m);

c = a coefficient depending on nature and temperature of soil (for clay $c = 0.0015$ and for sand $c = 0.003$).

(v) In the U.S.S.R. the following formula is used to calculate a general seepage loss rate (A5):

$$S = \frac{1.16}{Q} \times K \times q_r$$

where

S = loss as percentage of canal discharge per km of canal length;

Q = canal discharge (m³/sec);

K = saturated permeability (m/day);

q_r = reduced specific seepage loss, i.e., ratio of seepage velocity to saturated permeability of bed material.

(vi) Offengenden proposes the following equation for estimating seepage losses from earth canals or ditches (B9):

$$S = s \frac{Q \times l}{100} \text{ m}^3/\text{sec}$$

where

s = water losses per km of canal length (in percent);

Q = water flow (m³/sec).

l = length of canal (km);

S is calculated by the formula:

$$S = \frac{A}{Q^m}$$

A and m are empirical constants depending on soil permeability:

	Permeability		
	Low	Medium	High
A	0.70	1.90	3.40
m	0.30	0.40	0.50

It can be seen from the last equation that conveyance efficiency increases in a given canal as the volume of flow increases, and decreases rapidly with higher soil-permeability values.

Formulae to calculate seepage losses in periodically functioning canals have been developed by Offengenden (B9).

Analytical solutions

Analytical solutions of seepage problems related to irrigation canals have been presented by Dachler, Ernst, Hammad, Harr, Pawlowsky, Polubarinova-Kochina, Vedernikov, Wesseling, and others. Although most existing analytical solutions are accurate tools for seepage calculation, they are too numerous to be given here. (For detailed information, see B10 and B21 and references therein.) However, the most important of these are included in Bouwer's graphical solutions.

Graphical solutions derived from electrical analogy

The practical application of Bouwer's graphical solution for predicting seepage rates for a given canal means that the following factors governing the flow system must be known (B10): hydraulic conductivity (K) of the subsoil, geometry of the canal, and position of the groundwater table.

According to Bouwer, there are three basic conditions to which the multitude of natural profiles of soil hydraulic conductivity can be reduced for theoretical treatment of seepage flow systems:

Condition A: The soil in which the channel is embedded is uniform and underlain by more permeable (considered infinitely permeable) material.

Condition B: The soil in which the channel is embedded is uniform and underlain by less permeable (considered impermeable) material.

Condition C: The soil in which the channel is embedded is of much lower hydraulic conductivity than the original soil for a relatively short distance normal to the channel perimeter (clogged soil, semipermeable linings).

Seepage to a free-draining permeable layer in the subsoil is a special case of condition A; it is obtained by allowing the water table to be at or below the top of the permeable material. This condition is labelled A'.

The studies of canal seepage by resistance network analogue included analyses under the conditions A, B and A'. The geometry and symbols for canals under these three conditions are shown in Figure 6.

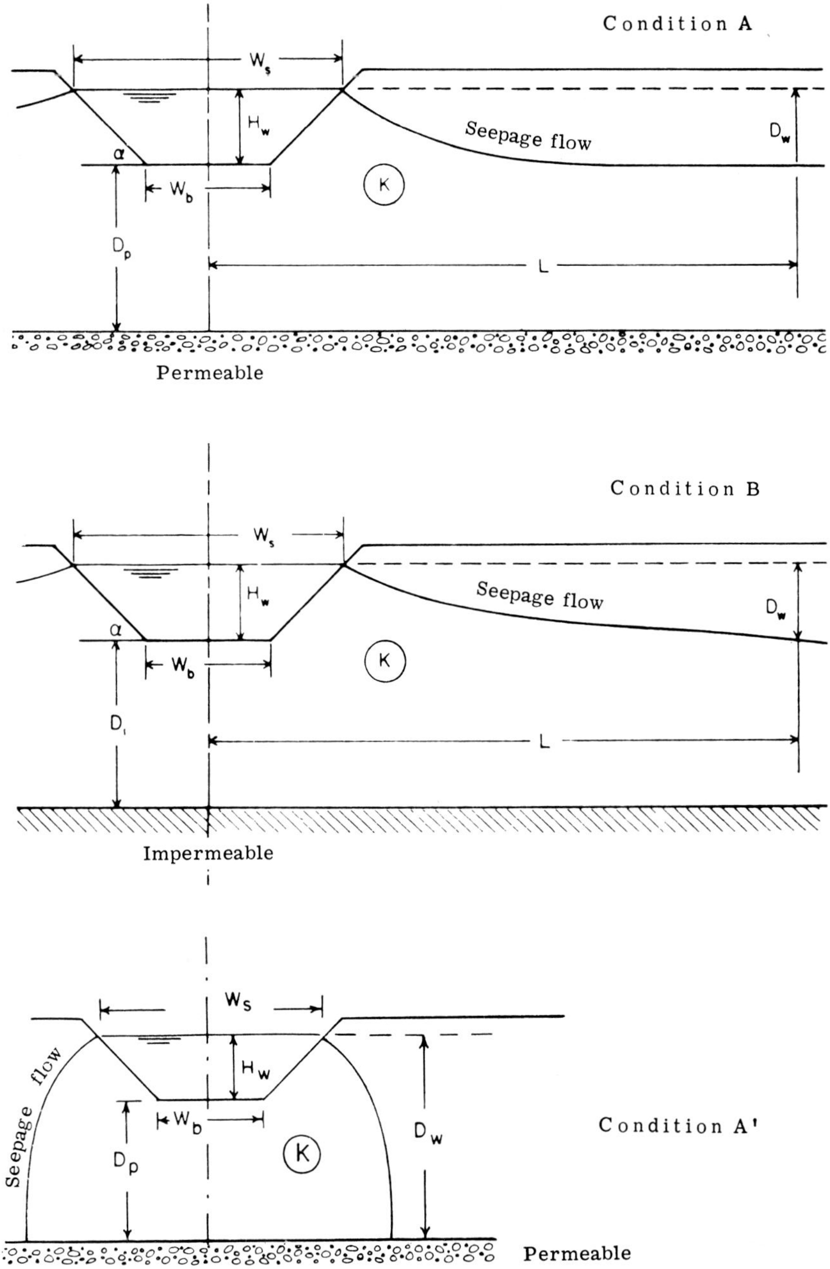

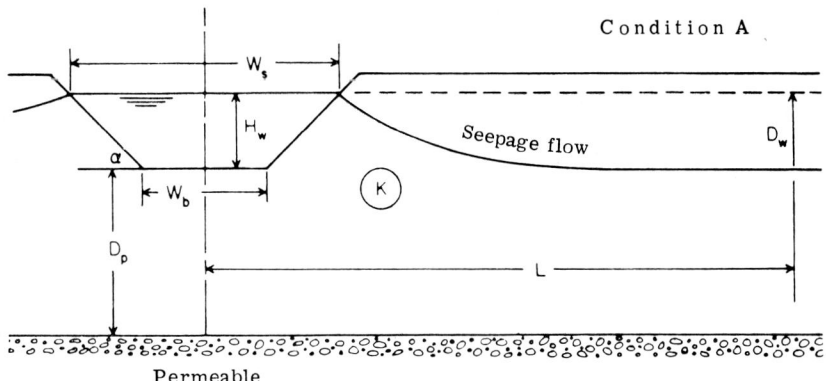

FIGURE 6. Geometry and symbols for canals under conditions A, B, and A′ (B10).

D_w is the head, which affects the seepage flow. For conditions A and B this is equal to the vertical distance between the free water surface and the horizontal water table. In the analogue the horizontal water table was simulated at a horizontal distance (L) of 10 times the bottom width (W_b) from the canal centre. For condition A' the effective D_w value is equal to $H_w + D_p$, although the actual depth of the water table may be greater than $H_w + D_p$. The analyses were performed for trapezoidal canals with 1:1 side slopes and three different water depths (expressed as H_w/W_b). The graphical solutions resulting from the resistance network analogue are shown in Figures 7, 8, 9.

Use of the graphs. The dimensionless value I_s/K can be obtained from the graphs. From this the seepage rate (q) per metre of canal per day is computed by using the expression

$$q = \frac{I_s}{K} \times K \times W_s.$$

To apply the graphs to canals of other shapes, W_b is computed from the actual values of W_s and H_w as if the canal were trapezoidal with a 1:1

FIGURE 7. Results of seepage analyses by electrical analogy for a trapezoidal canal ($H_w/W_b = 0.75$).

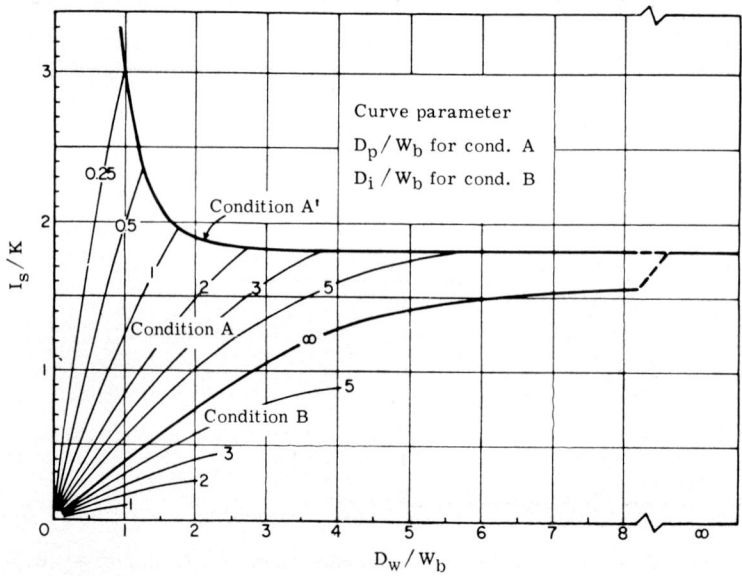

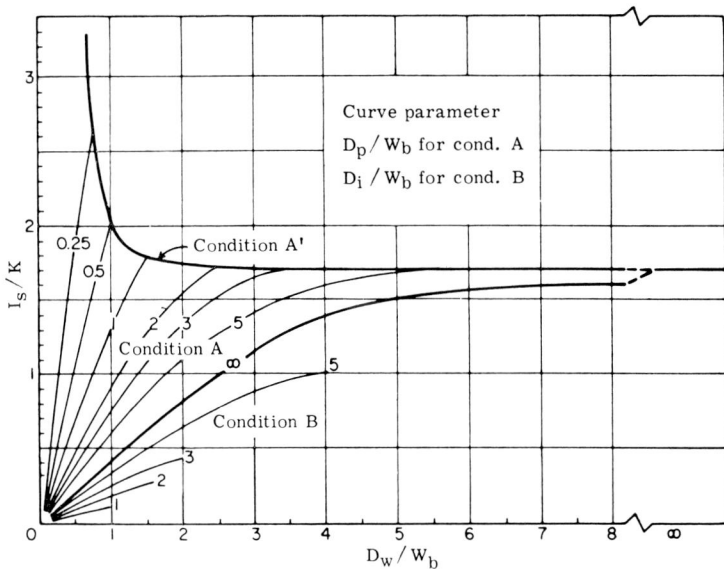

FIGURE 8. Results of seepage analyses by electrical analogy for a trapezoidal canal ($H_w/W_b = 0.50$).

FIGURE 9. Results of seepage analyses by electrical analogy for a trapezoidal canal ($H_w/W_b = 0.25$).

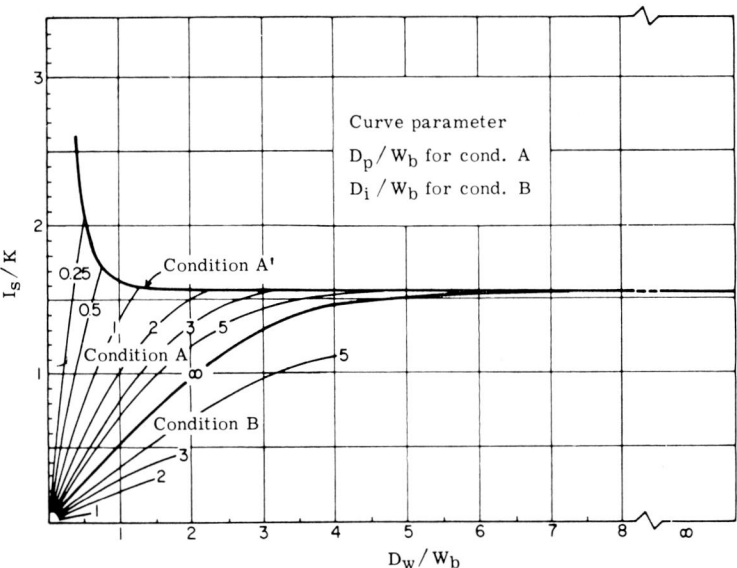

side slope, or the cross section can be replaced by the best-fitting trapezoidal cross section with a 1:1 side slope.

In practice, the underlying soil material is treated as "permeable" if its K value is 10 times greater than the K value of the soil above (condition A). Accordingly, the state of impermeable floor can be presumed if the hydraulic conductivity of the material is 10 times less than that of the soil above (condition B).

The graphs show that the influence of the permeable layer (condition A) on the seepage rate becomes rather small when it is more than 5 times the bottom width (W_b) below the canal. The closer to the canal bottom the permeable layer lies, the higher the seepage rate. The rate also increases with an increase in the water table depth (D_w).

For condition A' it is seen that the seepage rate remains almost constant for a wide range of depths of the permeable layer (= effective head D_w). When this depth becomes less than 3 times the canal depth (H_w) the seepage rate increases rapidly.

For condition B it is evident that the impermeable floor has a significant effect on seepage only if its distance below the canal (D_i) is less than 5 times the bottom width of the canal (W_b).

The following example illustrates the use of the graphs:

Given: The soil in which the canal is embedded is a sandy loam which can be considered uniform. The mean hydraulic conductivity is K = 0.50 m per day. This soil is underlain by material at least 10 times less permeable (considered impermeable). Thus condition B is given.

H_w = 0.75 m D_i = 5.00 m

W_b = 1.00 m D_w = 2.00 m (at distance 10 × W_b from canal centre)

W_s = 2.50 m

Side slope 1:1

From this:

D_i/W_b = 5:1 = 5

D_w/W_b = 2:1 = 2

H_w/W_b = 0.75:1 = 0.75

for which Figure 7 gives $I_s/K = 0.58$. Thus

$$q = \frac{I_s}{K} \times K \times W_s = 0.58 \times 0.50 \times 2.50 = 0.73 \text{ m}^3/\text{m of canal/day}.$$

METHODS FOR DETERMINING SOIL HYDRAULIC CONDUCTIVITY

The application of the solutions presented in the previous paragraphs toward predicting seepage rates for a given canal requires that the soil and boundary conditions governing the flow system be known. The extent and precision to which the conditions have to be determined depend on individual requirements. No general guidelines can be given. Of particular importance is the proper determination of the hydraulic conductivity (K) of the soils.

Table 8 shows that the permeability of natural earth material varies widely; for example, a clean gravel carries water at a rate of about one thousand million times that of an unweathered clay.

TABLE 8. — MAGNITUDE OF SPECIFIC HYDRAULIC CONDUCTIVITY

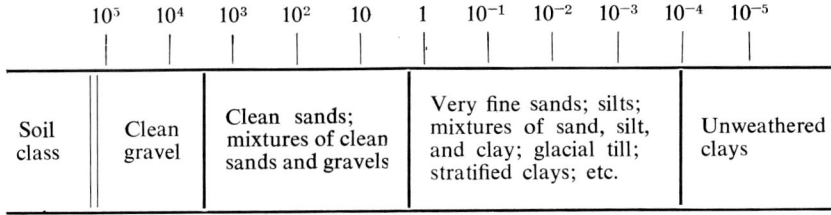

Soil class	Specific hydraulic conductivity (K) darcys (feet per year)			
	10^5 — 10^4	10^3 — 10^2 — 10 — 1	10^{-1} — 10^{-2} — 10^{-3}	10^{-4} — 10^{-5}
	Clean gravel	Clean sands; mixtures of clean sands and gravels	Very fine sands; silts; mixtures of sand, silt, and clay; glacial till; stratified clays; etc.	Unweathered clays

Hooghoudt quotes the following range of hydraulic conductivity values for the Netherlands in metres per day:

Sands = 0.1 for fine sands to 30 for coarse sands;

Clays = 0.01 to more than 30 (some clay soils have better permeability than coarse sandy soils);

Peats = 0.01 to more than 10.

The in-place field techniques for K measurement can be divided into those that measure K of the soil below the water table and those that measure K of the soil above or in the absence of a water table. The latter are of special interest in seepage prediction, because the soil in which the canal is, or will be, embedded is commonly not affected by a water table.

As was stated previously, the accurate calculation of seepage, including K measurements, is rather sophisticated and will in any case require careful preparation, including collection of the most recent technical information. Therefore, the accepted methods for K measurement are only briefly described here but with references to sources.

Methods of measuring hydraulic conductivity below the water table

The average hydraulic conductivity over a relatively large soil region can be obtained with pumping-test techniques established for groundwater investigations. (References B22, 36, 37, 38 and 42 provide detailed information.)

Faster and simpler ways of obtaining K are the auger-hole method, the piezometer method, the tube method and the multiple-well technique. K-values thus obtained are, however, confined closely to the quite shallow and small-diameter holes employed in these methods.

The auger-hole method gives the average permeability of the soil layers extending from the water table to a small distance (a few decimetres only) below the bottom of the hole (if in a permeable stratum). The radius of the column of soil for which the permeability is measured is about 30-50 cm.

The general principle of the auger-hole method, and with slight modification also of the piezometer and tube techniques, is simple:

(*a*) a hole is bored into the soil to a certain depth below the water table;

(*b*) a quantity of water is removed from the hole after equilibrium conditions have been reached;

(*c*) as the surrounding groundwater seeps into the hole again, the rate at which the water level rises in the hole is measured (usually in cm/ 10 seconds) and then converted by a suitable formula to the hydraulic conductivity of the soil.

Auger-holes are rarely bored deeper than 1 m below the water table. Diameters range from 4 to 8 cm. (For comprehensive information on the auger-hole method see B41 and 44, which contain graphs giving K for known rates of rise in water level.)

The piezometer method, which is particularly suitable for measuring hydraulic conductivity in stratified soils (B44), has been developed by Kirkham. A piezometer — an unperforated pipe with one sharpened end into which water can enter — is pressed down into the soil. By repeated boring and pressing over short distances the pipe can be placed without compressing the soil around, and without leaving openings around the pipe. At the desired depth the soil is bored out some centimetres below the pipe, thus making a cavity where the water can flow into the pipe. The rate of rise of the water level in the pipe is measured as with the auger-hole method (for example, by electric probe). The measured rate of rise can be transformed into a K-value by using an equation developed by Kirkham. (For detailed information see B24, 43 and 44.)

The tube method developed by Frevert and Kirkham is a special type of piezometer method which in essence measures the vertical permeability of the soil (B17).

In the multiple-well method, several wells a relatively short distance apart are used, and K is calculated from the flow system created by pumping from one well into another. The method was first proposed for two wells by Child (B17).

Methods of measuring hydraulic conductivity above the water table

Most irrigation canals are constructed in areas with a relatively low water table. Therefore, those methods developed for measuring K in soils above the water table are the most important for seepage investigations. The general principle of these methods is to wet a portion of the soil, preferably to positive soil-water pressures, and create in this wetted zone a flow system of known behaviour for the evaluation of K. Devices have been developed either for simple one-dimensional (i.e., vertical) flow or for axisymmetric flow. Vertical flow measurements refer to seepage from canals to deep groundwater tables, whereas axisymmetric flow measurements relate to seepage from canals to shallow groundwater tables.

Methods based on vertical flow measurements are the infiltrometer techniques and the air-entry permeameter. (For detailed information, see B10, 12, 13 and 14.) As described by Bouwer (B10): "The time required per test with the air-entry permeameter may range from 0.3 to 1 hour depending on the type and water content of the soil. Approximately 10 litres of water are required per test. The soil should be sufficiently dry to permit the development of a well-defined, easily detectable wet front. In its present form of construction, the air-entry permeameter is a surface device. Sub-surface measurements can be obtained by placing the device in the bottom of pits or trenches."

Two important methods based on axisymmetric flow systems are the well permeameter method and the double-tube method.

The well permeameter method. In the well permeameter method a hole 10 to 15 cm in diameter is bored in the soil to the desired depth (not less than 5 times the diameter), and the lower part of the hole is filled with coarse gravel. The top portion, in which the float arrangement is installed, is provided with a screen or casing to prevent the sides from caving in. Water is delivered to the hole from a calibrated tank which has a glass indicator tube, and the outlet is provided with a valve coupled to the float arrangement housed in the hole to maintain a constant water level in the hole.

The seepage rate is indicated by the rate of loss of water from the tank into the hole. The hydraulic conductivity can be calculated using formulae developed by the U.S. Bureau of Reclamation and published in its *Earth Manual* (F3).

The well permeameter method is elaborate, costly and time consuming because the test should be continued for several hours to attain equilibrium conditions and a large quantity of water is required, which may be difficult to arrange especially if the tests are to be carried out in remote locations.

Talsman (C87) concludes from field test comparisons that "hydraulic conductivity values measured by the well permeameter method were on the average 33 to 61 percent lower than the corresponding auger-hole values. These deviations are of the same order of magnitude as the combined reduction in hydraulic conductivity caused by puddling of the sides of the holes due to augering and subsequent closing of the pores around the hole by the infiltration process. The well permeameter method is considered of value in studies where a knowledge of the order of magnitude of the hydraulic conductivity is required."

The double-tube method. The double-tube method, a relatively recent development, is described by Bouwer (B10).

The apparatus consists of two concentric tubes which are inserted into an auger hole and covered by a lid with a standpipe for each tube (see Figure 10). Water levels are maintained at the top of the standpipes to create a zone of positive water pressure in the soil below the bottom of the hole. The hydraulic conductivity of this zone is evaluated from the reduction in the rate of the flow from the inner tube into the soil when the water pressure inside is allowed to become less than that outside the inner tube. This is done by stopping the water supply to the inner tube (closing valve B) and measuring the rate of fall of the water level in the standpipe on the inner tube while keeping the standpipe on the outer tube full to the top. This rate of fall is less than that obtained in a subsequent measurement in which the water level in the outer-tube standpipe is allowed to fall at the same rate (by manipulating valve C) as that in the inner-tube standpipe. The difference between the two rates of fall enables the calculation of K. Dimensionless factors evaluated by electric analogue are used in this calculation.

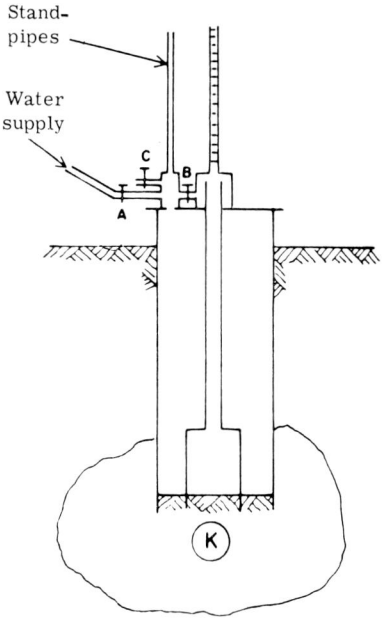

FIGURE 10. A double-tube apparatus (B10).

Although theoretically not limited by depth, the practical depth range of this method is approximately 0.5 to 5 cm. Depending on the type of soil and the depth of the hole, tests are usually completed one or two hours after the tubes are filled with water. Approximately 200 litres of water are required per test.

Laboratory methods of measuring permeability

Laboratory tests of soil permeability along the line of a proposed canal may be made on samples of either disturbed or undisturbed material. Experience has shown that the laboratory methods for permeability determinations are not so reliable as field methods; however, laboratory tests using undisturbed samples may be helpful for locating areas of relatively higher seepage (qualitatively, not quantitatively).

The Kopecki-ring method (B44) is a simple means of determining hydraulic conductivity of undisturbed samples. In the field, a thin-walled cylinder is pressed into the soil to obtain a soil sample with an essentially undisturbed structure. The soil protruding at the lower side is cut away. A screen is placed at the bottom of the sample to prevent loss of soil. A water head is fixed above the sample, which is kept on a constant level. The water which flows through the sample in a fixed time is measured in a graduated glass. A diagram of the apparatus used in the Kopecki-ring method is shown in Figure 11.

The hydraulic conductivity is calculated from the formula

$$K = 864 \, \frac{L}{h+L} \times \frac{F}{Q} \times \frac{\eta_t}{\eta_{10}}$$

where K = hydraulic conductivity (m/24 h);
L = thickness of soil sample (cm);
h = water head above sample (cm);
Q = discharge (cm³/sec);
F = inner cross section of the cylinder (cm²);
η_t = viscosity of the used water (poise);
η_{10} = viscosity of water at 10°C.

Values of η_t for different temperatures are:

Temperature (°C)	Viscosity (η_t: poise)	Temperature (°C)	Viscosity (η_t: poise)
0	0.01794	22	0.00961
2	0.01674	24	0.00916
4	0.01568	26	0.00875
6	0.01473	28	0.00836
8	0.01387	30	0.00800
10	0.01309	32	0.00767
12	0.01239	34	0.00736
14	0.01175	36	0.00706
16	0.01116	38	0.00679
18	0.01060	40	0.00654
20	0.01009		

CONCLUSIONS

The decision whether or not to line a canal essentially depends on the permeability of the soil in which the canal is to be excavated. In many practical cases this decision can be reached from visual observations of the soil, provided that it is of a type which is obviously very pervious or impervious. When permeability is in doubt the decision may be reached either by applying comparative seepage data or by calculating seepage in conjunction with the determination of the hydraulic conductivity by field tests. These tests and calculations are elaborate, time consuming and expensive if accurate results are to be obtained. Unless a precise prediction of seepage is of particular importance for the decision to line, and unless the investigations and calculations are conducted by a specialist, the result obtained may not warrant the expenditure of time and money.

In many cases it may be more advantageous to adopt available seepage data from nearby unlined canals or to undertake seepage measurements as described above. Seepage data for different soils and lining materials provided in Table 2 serve as a guide where no other data are available and where investigations are extremely difficult or too costly.

Because sedimentation and other "aging" processes causing reduced seepage rates are not taken into account, the seepage rates evaluated with the above methods should be considered the maximum that can be expected. Such rates may occur when the canal begins operating or after it has been cleaned.

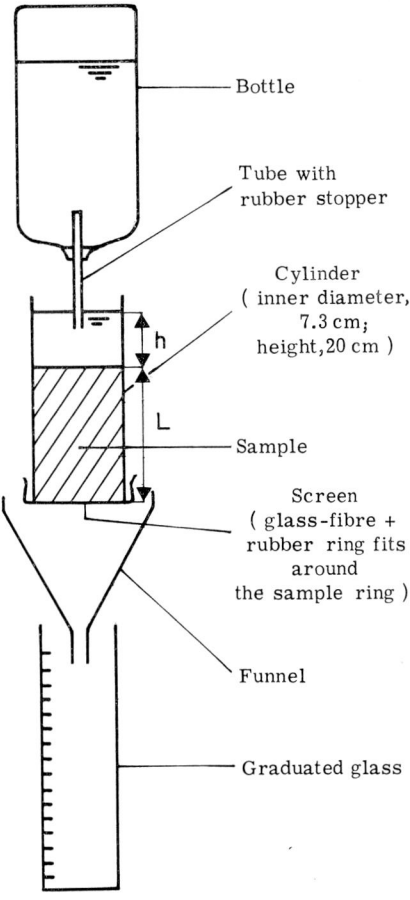

FIGURE 11. Equipment used for the Kopecki-ring method of determining hydraulic conductivity in undisturbed samples (B44).

37

Measuring seepage from existing canals

The objectives of post-construction seepage measurements may be:
(a) to determine seepage losses from unlined canals and to locate reaches with excess seepage as a basis for lining considerations;
(b) to check seepage losses in completed reaches of a canal system under construction with the aim of predicting seepage rates in the uncompleted parts of the system and adapting the design to the findings. This could involve lining where this had not been foreseen, and vice-versa, changing the type of lining or modifying the size of the canal cross section and structures according to actual flow rates;
(c) to record seepage rates on lined or unlined canals as comparative data for the planning and design of other irrigation projects;
(d) to determine the exact amount of water conveyed in the canal system in order to operate the system properly.

Currently accepted methods of measuring the quantity of water lost by seepage from existing canals are limited to inflow-outflow, ponding, and seepage meter determinations. Special methods are the use of tracers, electrical logging or resistivity measurement, piezometric surveys and remote sensing. These special methods are essentially limited to qualitative indication of the distribution of seepage along the canal.

Each method has advantages and limitations. No single method is adaptable to all conditions encountered in the field. Ponding may be the most accurate and dependable method.

INFLOW-OUTFLOW METHOD

The inflow-outflow method consists in measuring the water that flows into and out of the section of irrigation canal being studied. The difference between the quantities of water flowing into and out of the canal reach is attributed to seepage.

If seepage losses are small, evaporation and precipitation must also be taken into consideration even though those factors generally have no significant effect on seepage loss. Published data on evaporation are sufficiently accurate for this purpose, but are not available for many areas. The stage of the canal should be constant during test periods in order to eliminate the effect of bank and canal storage. All diversions and leaks within the test reach must be measured accurately; also any inflow into the canal from surrounding areas must be taken into account. Leaks that cannot be eliminated are best measured volumetrically with a calibrated device.

The following techniques are usually employed for discharge measurements in the inflow-outflow method.

Current meter

This is probably the most practical. Any standard type of meter in good condition and accurately calibrated may be used. Two instruments and two people are required. Errors due to personal observations and instruments should be compensated by changing the instruments and observers between the head and tail discharge measuring sites. The two-point method and the six-tenths depth method are most used in canal work for determining mean velocities. The two-point method is recommended for larger canals. It consists in measuring the velocity at 0.2 and then at 0.8 of the depth from the water surface, and using the average of the two measurements. The accuracy obtainable with this method is high.

The six-tenths depth method consists in measuring the velocity at 0.6 of the depth from the water surface, and is generally used for shallow flows where the method previously described is not applicable. The procedure gives fairly satisfactory results. (Detailed information may be found in B6, 7, 34 and 35.)

Flumes, drops, weirs, orifices

Properly constructed flumes or sharp-crested weirs equipped with automatic recording gauges provide the most accurate measurement of water volume. If automatic gauges are not available, observations of the volume of water passing through such structures spaced at suitable intervals, with proper time lags between observations, are reasonably accurate. (Detailed information is found in B6, 7, 34 and 35.) Inflow-outflow measurements can be made fairly easily and interfere little with the operation of the canal. They are only as good as the accuracy of the method and the individuals making the measurements. This method is advantageous when seepage losses are to be measured in long canal reaches, with few diversions, to show the general efficiency of the canal with regard to seepage. It is also sufficiently accurate in canals with large seepage losses; however, great care has to be taken to seal off all outlets and offtakes so that the difference between inflow and outflow does not include leakage through such structures but reflects actual seepage losses.

The best accuracy over a wide range of discharge can be obtained with the V-notch weir shown in Figure 12. Construction details and rating tables are found in a number of texts — for example, in the USBR *Measurement Manual* or the FAO/ICID *Handbook on Small Hydraulic Structures.*

FIGURE 12. Measuring flow with a 90° V-notch weir, western Macedonia, Greece.

PONDING METHOD

The ponding method consists in measuring the rate of drop in a pool formed in the canal reach being tested and in computing from this the seepage rate and ratio of the water surface area of the pool to the wetted area of the section. Since observations can be made accurately, the results should be a good indication of the average loss from the reach with the reservation that the still water in the pool may seep out at a different rate than the flowing water in the canal. This may be caused by the sealing effect of suspended material settling in the still water; by the growth of algae or fungus on the wetted perimeter, especially on lined canals; and by a change in the groundwater table when the canal upstream and downstream from the ponded reach is empty.

To isolate a reach of a canal for ponding tests, watertight dikes or bulkheads have to be built. Whenever possible, existing structures such as weirs and regulators should be utilized for this purpose.

To eliminate the effect of wind, the rate of drop should be measured at each end of the pool and averaged. Staff or hook gauges attached to existing structures or stakes driven into the canal bed should be used. All structural leaks should be carefully measured, and since the testing may take considerable time, evaporation and rainfall should be recorded so that the drop in water surface can be corrected accordingly.

The following formula is suggested for computing the rate of seepage:

$$S = \frac{W(d_1 - d_2) \times L}{PL}$$

where

S = average seepage in m³/m²/24 h (or ft³/ft²/24 h) over distance L;

W = average width of water surface of the ponded reach;

d_1 = depth of water at beginning of measurement;

d_2 = depth of water after 24 hours;

P = averaged wetted perimeter;

L = length of the canal reach.

A modification of the ponding method consists in adding water to the pond to maintain a constant water surface elevation. The accurately measured volume of added water is considered equal to the total losses, and the elapsed time establishes the rate of loss.

The ponding method provides an accurate means of measuring seepage losses and is especially suitable when they are small (e.g., in lined canals). The results from this method are generally used as the standard of comparison for other methods of seepage measurement.

In addition to seepage measurement, the pond may be used to observe the structural behaviour of a lining under drawdown.

There are some disadvantages to the ponding method:

— ponding tests require interruption of the normal working of the canal;
— the construction of dams to form the pool and their removal afterwards are relatively costly. The method can therefore be used only when the importance of the test warrants fairly large expenditures;
— quite large amounts of water are required to fill the pool and also during the tests to compensate the drop in water level. If gates are installed in the dam for filling the pool, they must be watertight;
— although the ponding method gives accurate figures for the total seepage from the pool, it does not show the variation in rates from different parts of the pool.

Examples of ponding tests

A ponding test carried out on the Right Main Canal of the Chambal Commanded Area, Rajasthan (B26) was made to gain more accurate data on seepage than had been achieved by previous inflow-outflow measurements and to evaluate the role of seepage in the waterlogging process. In this case inflow-outflow measurements with current meters indicated that the errors were mainly greater than the seepage losses themselves.

A 17 000 ft (5 185 m) long reach (design discharge = about 4 500 cusecs) was marked off by cross regulators, the upstream one of which was also used to control the water supply to the reach. The regulators were sealed by constructing a wall of clay bags over the entire cross section. Seven days before the test started, the control reach was flooded and the water kept near the designed full supply level to allow the bottom and embankments to become saturated. Then the water supply was stopped and the rate of decrease in water level was regularly measured at the head, tail and two intermediate sections for 48 hours. The reach was then refilled and the observations repeated for another 48 hours. The leakage losses through the tail regulator were measured at the same time.

The mean wetted perimeter was determined by surveys at 15 cross sections. The total fall was $7\frac{1}{4}$ in (18.4 cm) during the first test and $7\frac{1}{2}$ in (19 cm) during the second. From the observed data the following results were calculated:

Total loss in the reach:	9.96	cusecs
Evaporation losses:	0.96	cusec
Leakage through lower regulator:	0.25	cusec
Seepage losses:	8.75	cusecs
Total wetted area:	2.84	million ft^2
Unit seepage losses:	3.08	cusecs/million ft^2
Design discharge:	4 500	cusecs
Total seepage losses relative to the design discharge of the reach:	0.22	percent

(For descriptions and evaluation of ponding tests, see also B15, 17.)

Figures 13a,b show two ponding tests conducted in western Macedonia, Greece, using ordinary jute sacks filled with a loamy soil. Excellent watertightness was obtained when layers of cohesive soil were placed between the sack layers. The filled sacks were re-used several times.

FIGURES 13a, b. Ponding tests with ordinary jute sacks in distribution ditches, western Macedonia, Greece.

SEEPAGE METER METHOD

The seepage meter is a modified version of a constant-head permeameter developed for use under water. Various types of seepage meters have been developed. The two most important will be described here.

Seepage meter with submerged flexible water bag

This is perhaps the simplest and cheapest device as regards construction as well as operation. It consists of a watertight seepage cup connected by a hose to a flexible (plastic) water bag floating on the water surface (Figure 14).

Water flows from the bag into the cup, where it seeps through the canal subgrade area isolated by the cup. By keeping the water bag submerged, it will adapt itself to the shrinking volume so that the heads on the areas within and outside the cup are equal. The seepage rate is computed from the weight of water lost in a known period of time and the area covered by the meter.

As may be seen from Figure 14, this type of device is fairly simple and can be constructed easily. The length of the handle and hose should be chosen according to given local conditions. The cylinder should be pushed only a small distance into the subgrade in order to avoid, as far as possible, disturbance of the existing soil texture. While submerging and pushing the cylinder into the subsoil, the hose is kept open at its upper end to allow air and excess water from the cylinder to escape. When equilibrium is achieved, the hose is connected to the plastic bag containing the weighed quantity of water. The design shown is only an example, and it may be changed and improved according to local possibilities. For example, a valve for air and water release can be installed on the cylinder and operated above the water surface by means of a bar attached to the handle.

The accuracy of measurement depends largely on the maintaining of an exact balance of pressure on the canal bottom inside and outside the meter. Analyses undertaken to study errors due to pressure differences indicate that, when seepage losses are low, a small head difference of 1 cm or so could cause errors in the measured seepage approaching the magnitude of the seepage. For high seepage rates, however, such pressure differences cause only a small error in the measurement (B4).

Falling-head seepage meter

It is difficult to maintain a head inside the seepage meter which is exactly equal to the head in the canal; therefore, a falling-head technique, whereby the seepage meter is connected to a falling-level reservoir, is used. A sketch of this type of meter is shown in Figure 15.

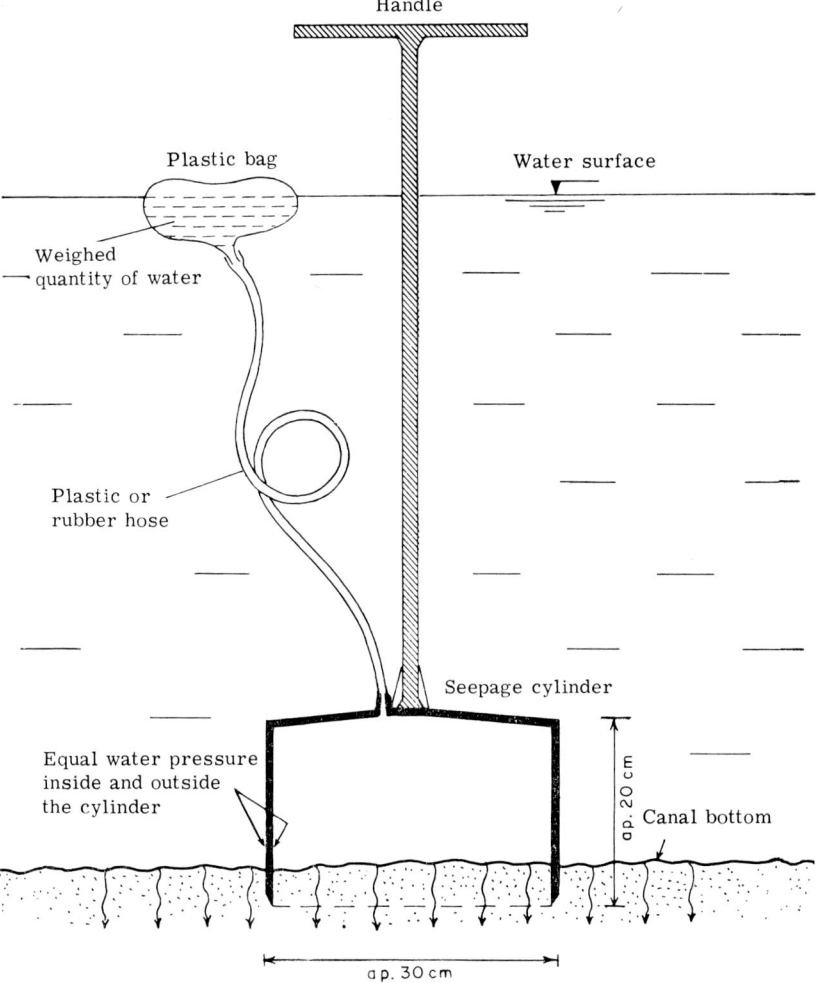

FIGURE 14. Seepage meter with submerged plastic bag.

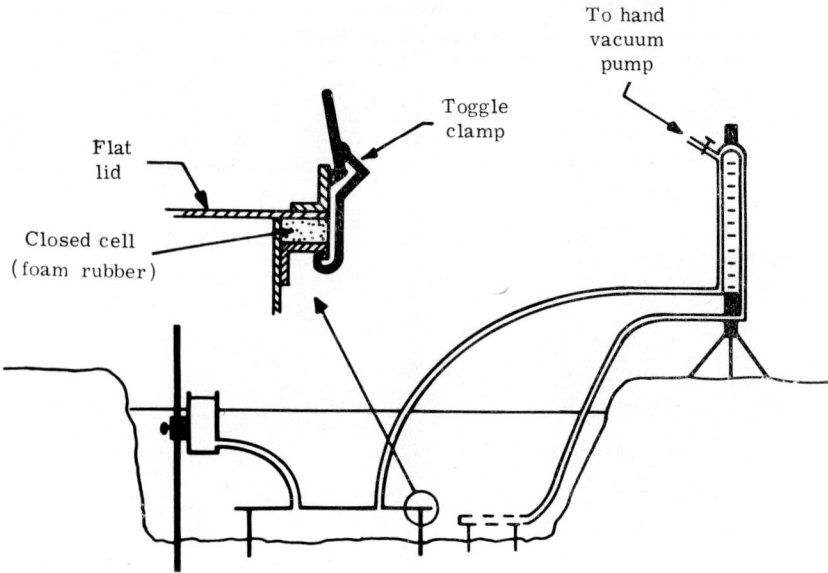

FIGURE 15. Falling-head seepage meter in a canal (USDA).

Before the seepage is measured, the water level in the reservoir is raised an inch or so above the water surface in the canal. The subsequent fall of the water level in the reservoir is measured by means of a vacuum inverted U-tube manometer placed on the bank of the canal. One leg of the manometer is connected to the seepage meter interior, and the other to the free water in the canal. The water level in the manometer tube connected to the seepage meter will fall, whereas that in the tube connected to the free water surface in the canal will rise. At any time, however, the difference between the water levels in the manometer tubes will be equal to the difference in pressure between the seepage meter and the canal, even if the water level in the canal is fluctuating during the time of the measurements.

Field measurements consist in taking timed readings of water levels in the manometer. From these data, seepage can be calculated graphically or analytically with a falling-head equation (Bouwer and Rice, B11). The graphical procedure is the easier. It consists in plotting the manometer and time readings on graph paper. At the point of intersection of the two resulting curves, the pressure inside the meter equals that outside in the canal, and the angle between the curves at their point of intersection can be converted into the seepage from the meter.

The procedure of graphical solution is shown in the following example:

SEEPAGE METER TEST

Date: _22 April 1970_   Location: _Henares Canal_

Operators: _G. Castanon - D. Garcia_

Water depth: _0.75 m_   Bottom width of canal: _2.00 m_   Velocity: _0.50 m/sec_

Bottom material: _approximately 10 cm silt on silty loam_

Diameter of bell: _30 cm_   Penetration of bell: _4 cm_   Diameter of reservoir: _5 cm_

WATER LEVEL IN MANOMETER TUBES

Time (min)	Seepage meter interior (cm)	Free water (cm)
0	10.0	8.0
½	9.5	8.1
1	9.2	8.3
2	8.5	8.5
3	7.9	8.7
4	7.2	9.0

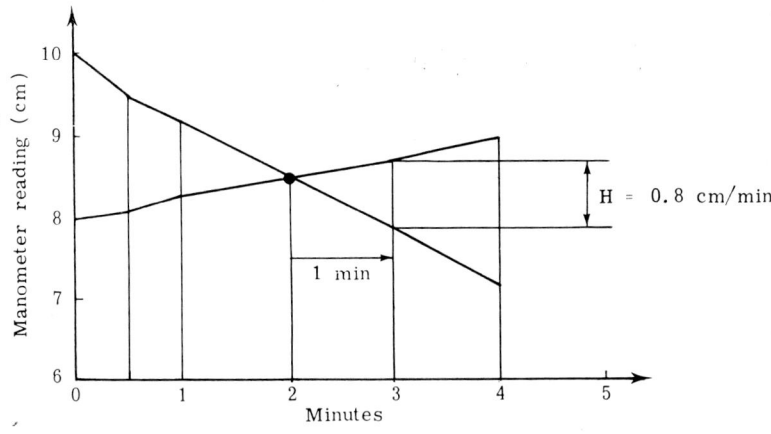

After plotting the recorded data as shown above, the vertical distance H at a predetermined distance of 1 minute from the point of intersection is measured.

With the given values

$$H = 0.8 \text{ cm/min},$$
$$R_v = \text{radius of reservoir} = 5 \text{ cm},$$
$$R_c = \text{radius of the cylinder} = 15,$$

the seepage rate q_s is computed with the following formula (B11):

$$q_s = H \frac{R_v^2}{R_c^2} = 0.8 \frac{25}{225} = 0.089 \text{ cm/min}$$

$$= 1.28 \text{ m}/24 \text{ h}.$$

Application considerations

Seepage meters are, in principle, suitable for measuring local seepage rates in canals or ponds, but they can obviously be used only in unlined or earth-lined canals. They are quickly and easily installed and give reasonably satisfactory results for the conditions at the test site. They are particularly useful for locating short sections of a canal where seepage is excessive and where lining should be considered.

Seepage meters should be installed with the least possible disturbance of the bed material. Disturbance of the soil during insertion of the meter can cause indicated seepage rates to be higher than actual. Comparison of the falling-head meter method with the ponding method has shown 30 percent higher seepage rates for the seepage meter (B15). Warnick (B4) showed that, with a constant-head meter, pushing the seepage bell into the bottom of the canal caused measured seepage rates to be 23 percent greater than rates achieved by independent means of ponding. Gently stepping on the meter to position it did not change this relationship, whereas hammering on the seepage bell increased the meter rate over the control ponding rate by as much as 58 percent.

Seepage meters cannot be used in very gravelly soil because of the difficulty of forcing the bell into the bed of the canal; in sandy soil it is likely to be washed away by the current.

There is no general rule for estimating the number of seepage meter tests required to obtain a reasonable average value of the seepage rate from a section of canal, since this depends on local conditions. A formula restricted to certain preconditions has been proposed by Brockway and Worstell (B15). In developing the formula an average of 66 seepage meter tests were made in a 0.8 km reach of a 7.5 to 9 m wide canal, equal to about one test per 100 m². Two experienced men performed about 40 tests per day. Costs for obtaining an estimate of the seepage rate were about $300 per mile. For an equal length of test reach on the same canal a ponding test cost $1 500.

When using the falling-head meter, water depths are limited to less than 60 cm, even though the operating depth of the canal may be much greater. The advantages of its ease of operation and apparent accuracy at relatively shallow depths may be less advantageous at greater depths of water. Further studies are under way to develop a meter based on the variable-head principle which can be used efficiently in an operating canal (B15). However, seepage meters of the constant-head type can be used during normal canal operations, but it is difficult to obtain accurate results with this device when seepage losses are low.

SPECIAL METHODS

These methods refer to tracing and detecting seepage and its distribution along a canal, with the aim of locating sections with excess seepage (see B3, 4, 19, 21).

Tracer method

Fluorescent dyes and radioactive isotopes have been experimented with as seepage detectors. Some ten years ago they were considered promising, but little information is available on their actual applicability in the field.

Electrical logging or resistivity measurement method

Resistivity measurements can be used as the basis for estimating seepage or serve as a qualitative indicator of seepage because the electrical resistance of the soil varies with the water salt content. The technique consists in obtaining a continuous record on a strip chart of the variations in electrical resistance and in natural or self-voltage of material from point to point along the bottom or sides of the canal.

The natural or self-voltage of the material is believed to be induced by slow movement of water through fine-grained material and can be measured by use of very sensitive equipment. If the natural voltage in the material surrounding the canal shows little variation from point to point, the canal section is watertight or there is little seepage occurring. In contrast, if the natural voltage through the surrounding material changes rapidly at adjacent points along the canal, such a reach may have appreciable seepage.

The measurements are made by an instrumented truck moving along the canal banks dragging flat electrodes along the canal side or bed. These electrodes are connected to a source of alternating current and measurement equipment consisting of a chart recorder mounted in the instrument truck. Sixteen miles of canal have been logged in 8 hours.

Investigations were carried out by the U.S. Bureau of Reclamation to check the accuracy of this method by comparing the results with those of other methods — ponding, seepage meter, etc. — and it was found that the technique of electrical logging provides a considerable degree of verification, but some limiting factors have to be overcome.

Dhillon (B19) states that this technique has great potential importance as it provides a cheap and quick method for locating leakage zones in both lined and unlined canals.

Piezometric surveys

Observation of water levels in a series of piezometer tubes located at a right angle to the centre line of a canal provides data to determine the flow lines and equipotential lines of seepage water. The amount of seepage can then be calculated when the permeability of the soil is determined. This is a laborious process, but an experienced technician can estimate the amount of seepage occurring from a canal by studying the piezometric observations of variation in the water table. Evaporation losses from a water table located at a depth greater than 1.50 m (5 ft) are negligible and other outflow from the groundwater to the natural drainage can be accounted for. The rise in water table when a canal is put back into operation after a long closure also gives an idea of the extent of seepage.

Remote sensing

This term has been associated with any kind of data recording by a sensor which measures energy emitted or reflected by objects located at some distance from it. Usually, sensors are carried by aircraft.

From about a dozen types of sensors available at present, infra-red scanners and radar appear to be the most suitable remote sensors to supplement aerial photography in the detection and evaluation of seepage problems. With modifications and more precise calibration these devices may sometimes provide information for estimating the rate of seepage. Extensive research on the application of remote sensing in various field of agriculture, including seepage problems, is being carried on.[1]

---

[1] Information can be obtained from the National Aeronautics and Space Administration (NASA), Houston, Texas, U.S.A.

3. DESIGN AND CONSTRUCTION

Hard-surface linings

This category includes all exposed linings constructed of cement concrete, mortar, soil-cement, asphaltic materials, brick and stone.

GENERAL DESIGN CONSIDERATIONS

Cross section

Since the cost of hard-surface linings is usually high, the section with the smallest perimeter for a given area is the most economical. A semicircle has the smallest perimeter for a given area but is not practical as the top portions of the sides are too steep. The steepest satisfactory side slope for most canals from both construction and maintenance viewpoints is 1.5 to 1. Steeper slopes may be used on small canals where the soil materials remain stable.

Major limitations to the steepness of hard-surface linings are slippage of the lining and soil stability. Slippage may be caused by insufficient friction between the lining and the subgrade in combination with effects on external hydrostatic pressure (drawdown).

Canals provided with a hard-surface lining are usually designed with a finished bed-width to water-depth ratio of 1 to 2. Small canals normally have a ratio of 1, while the ratio for large canals may exceed 2. Some large brick-lined canals in India and Pakistan have ratios up to 10.

Designs for a canal should specify freeboards of sufficient height to prevent overtopping of the banks during sudden rises in water level. Adequate freeboards in hard-surface lined canals depend on the size of the canal, conditions of flow, curvature of alignment, entry of storm water into the canal, wind and wave action, increase in flow resulting from error at diversion, variation in the friction coefficient, accumulation of silt, and anticipated method of operation.

The normal freeboard for hard-surface linings ranges from 15 cm (6 in) for small canals to over 60 cm (24 in) for larger ones. The height of the canal bank above the top of the lining usually ranges from 30 to 60 cm (1 to 2 ft), depending on the size of the canal and local conditions. Figure 16 may be used as a guide.

Subgrade

A primary perequisite to the success of most hard-surface linings is a firm foundation in order to reduce the amount of cracking and the danger of failure due to settlement of the subgrade. Undisturbed soils are often satisfactory as a foundation for lining without further treatment.

Natural in-place soils of low density should be thoroughly compacted or removed and replaced with suitable material. Because they expand when wet, clays are usually hazardous to hard-surface linings and should be avoided.

If it is necessary to place a concrete or other rigid type of lining on expansive clay, there are several ways of reducing or controlling the probable damage. Clays vary so much in character that the pressure required to prevent expansion may be less than 0.07 kg/cm^2 (1 lb/in^2) in some types and as much as 10.5 kg/cm^2 (150 lb/in^2) in others. If the clay encountered can be controlled by loading the surface with non-expansive compacted soil, lining can be placed on this loaded subgrade and satisfactory service obtained. Similarly, if the expansive clay is a thin layer in an otherwise suitable subgrade, it has been found fairly effective to overexcavate the canal and replace the clay with gravel. Excavation to a depth of at least 60 cm has been the practice to date, but the depth and type of clay will influence the amount of excavation required.

Where overexcavation is prohibitive, alignment changes may be the best solution. Where neither excavation nor alignment changes are possible, an expensive method of chemical treatment of the clay, usually with lime, has been used successfully. The additional provision of horizontal construction joints in concrete linings at 0.2 m or 0.5 m from the bottom of the side slopes has prevented failures in some instances.

Gypsum soil is another hazard to hard-surface linings. Water in contact with gypsum will dissolve salts in the soil, in time creating cavities which may result in canal breaks and serious damage (Figures 17, 18). A totally waterproof lining would prevent most failures, but action by rainwater may also cause trouble. Successful solutions include making the lining waterproof by applying a cement-plaster coating over the existing concrete lining or placing a compacted layer of selected clay material under the concrete lining. Various drainage systems are also used to rapidly remove water that gets under the lining (A3).

Most difficulties occur in embankment sections. One of the most satisfactory solutions is to replace the embankment by a reinforced flume or

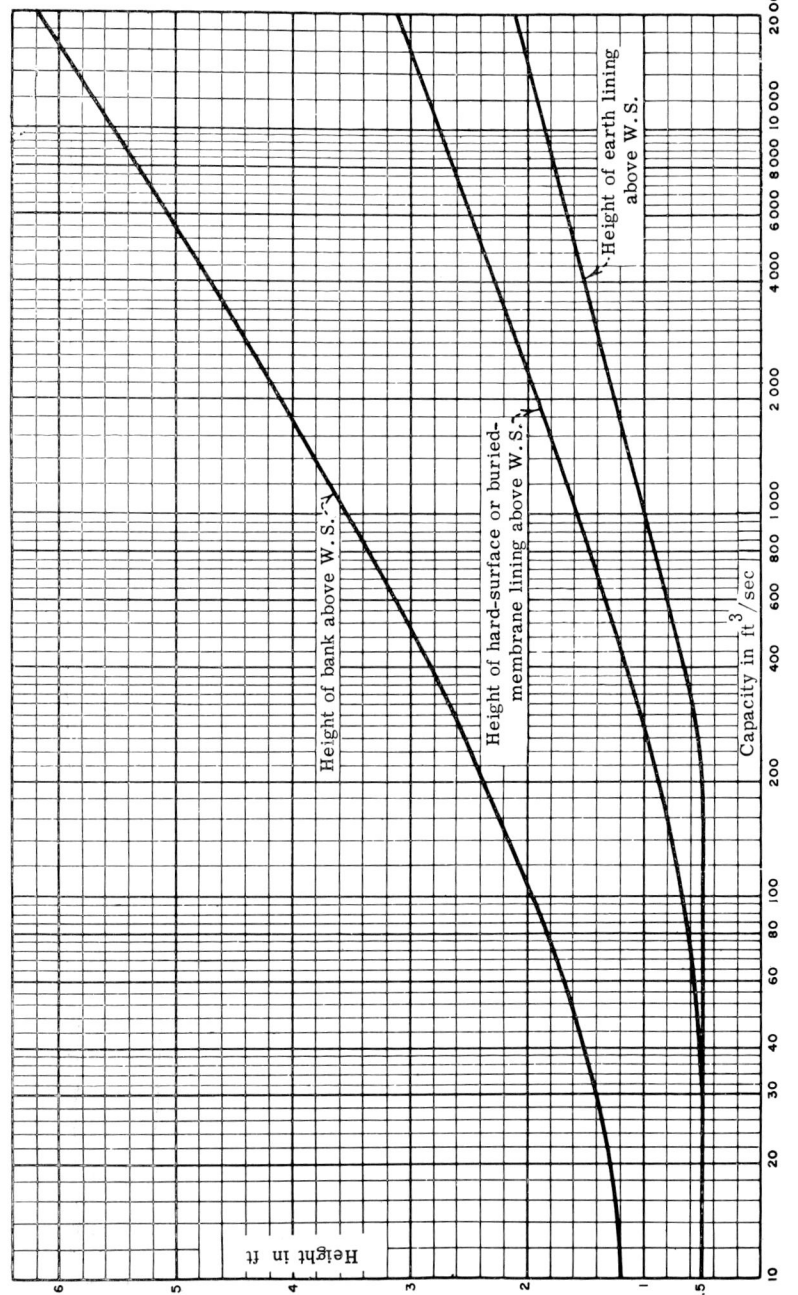

FIGURE 16. Bank height for canals and freeboard with hard-surface, buried membrane, and earth linings (USBR).

FIGURES 17-18. Views of a breakdown in the Monegros Canal, Spain, 1954 (F.H. Herrero, A3).

aqueduct set on piers or pedestals. This design is highly satisfactory but costly. Ideally, the alignment selected should avoid gypsum soils.

In concrete, mortar or brick masonry linings it is important that the subgrade be thoroughly moistened before placing the lining. Alternatively, the subgrade may be waterproofed with oil paper, crude oil or other appropriate and cheap temporary sealants.

In brick lining it is common practice to leave the compacted subgrade section projecting into the final section for, say, 0.5 m and cut this overhang just prior to lining operations.

Subgrade sterilization

Weeds are a potential hazard not only to earth or membrane linings, but also to asphalt-type linings, which they penetrate and, when dead, leave openings through which water can leak. Treatment of the subgrade with a soil sterilant is advisable when such linings are to be placed in areas already weed infested or in old canals where such weeds as tules, cat's-tails or willows are growing. Sodium chlorate as a 5 percent solution in water (2 l/m^2) is reported to be a satisfactory sterilizer (A15). Boron compounds, principally borax and boric acid, are favoured for use in conjunction with chlorates in sterilant treatments because the borates tend to leach more slowly. The U.S. Bureau of Reclamation recommended a water solution of polyborchlorate applied by spraying directly on the subgrade prior to placing the lining. Adequate sterilization is normally achieved by using an equivalent of 270 grams of powdered polyborchlorate per square metre of subgrade (A1).

Embankments

For most hard-surface linings the canal embankment must be compacted at least to the height of the lining. The top width of the compacted embankment (T in Figure 19) varies with size and location of the canal, type of lining, and other pertinent factors, but is usually 0.60 to 1.20 m (2 to 4 ft) for canals having a maximum capacity of about 3 m^3/sec (100 ft^3/sec) and 2.00 to 2.50 m (6 to 8 ft) for larger canals.

Loose material is placed over and outside the compacted embankment to provide space for operating roads and additional stability.

Unsuitable material should be stripped from beneath compacted embankments. Specifications for compacted embankments should require that after the necessary stripping has been done, the entire surface of the subgrade of compacted embankment be ploughed thoroughly to a depth of not less than 15 cm, moistened and compacted. The materials used in the embankment should have a specified moisture content and be compacted to a specified density in layers not more than 10 cm thick after compaction. The dry density of the soil fraction in the compacted material should not be less than 95 percent of the laboratory maximum density as determined by the Proctor method.[1] The maximum dry density and optimum moisture content for Proctor compaction of various standard soils are shown in Table 15 (see page 117).

---

[1] Specified in standard ASTM/D 558 (American Society for Testing Materials).

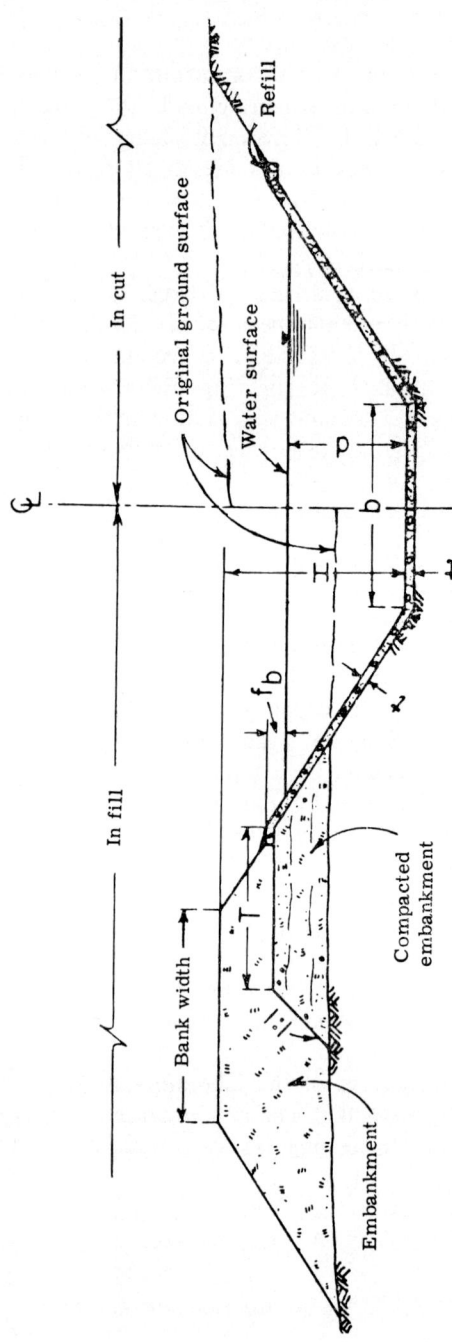

FIGURE 19. Typical cross section of a hard-surface-lined canal.

           a   *b*

FIGURE 20. Collapse of concrete lining caused by (*a*) lack of compaction and (*b*) undersized supporting embankments.

If Proctor-test equipment is not available on small work sites, the optimum moisture content can be approximately determined by adding water until the material can be formed manually into a compact ball. No water should be squeezed out and the material should maintain its dense structure after opening the fist. The compaction of loose soils in cut sections or of soil replacing unsuitable subgrade materials, or soil for backfill, should meet the same requirements for density.

Figures 20*a,b* show concrete lining which has deteriorated badly owing to lack of compaction and undersized supporting embankments.

It may occasionally be necessary to construct a hard-surface-lined canal in an area where the groundwater lies against or is likely to rise above the bottom of the lining. Such groundwater may expose the canal lining to external hydrostatic pressure when the canal is empty or running at low levels. Such pressure has frequently resulted in uplift and damage of the lining. For economic reasons, hard-surface linings should not be built to withstand excessive external pressure; instead, drainage should be provided underneath or alongside the canal to relieve any such pressure. Drainage is also an effective means of protection against frost heave.

Drainage

Occasionally, it may be necessary to construct a hard-surface-lined canal in an area where the groundwater is likely to rise above the bottom of the lining. In such cases drains must be provided underneath or alongside the canal to relieve any hydrostatic pressure which might cause uplift of the lining when the canal is empty.

There are two common types of artificial drainage installations. One consists of 10 or 15 cm (4 or 6 inch) till placed in gravel-filled trenches along one or both toes of the inside slopes. These longitudinal drains are either connected to transverse drains, which discharge the water below the canal or to pump pits, or extend through the lining and connect to outlet boxes on the floor of the canal. The outlet boxes are equipped with one-way flap valves or weep holes which relieve any external pressure that is greater than the water pressure on the upper surface of the canal base, but prevent backflow.

The second type consists of a permeable gravel blanket of selected material or sand and gravel pockets drained into the canal at frequent intervals (3 to 6 m) by flap valves in the invert. A drawing of a flap valve for use without tile pipe and in a fine gravel and sand subgrade is shown in Figure 21. Both the tile pipe system and the unconnected flap valve must be encased in an envelope which will prevent piping of subgrade material into the pipe or through the valve.

The concrete-lined canals in the Kangsabati Project, India, were provided with weep holes for each 2.80 m^2 (30 ft^2).

Water velocities

Flow velocities up to 12 m/sec are sometimes cited in the literature as permissible for concrete canals without properly indicating the lining characteristics and purpose of the canal. For irrigation canals it is advised to follow the USBR recommendation that velocities in unreinforced concrete linings should not exceed 2.5 m/sec (8 ft/sec) to avoid the possibility of lifting. This occurs when the velocity head is converted to pressure head through a crack that slopes upstream or into the current. A mathematical check using a Manning's *n*, which is 0.003 less than the design *n* used for the lining, is also recommended to ensure that the depth of flow does not approach critical depth closely enough to develop standing waves. Sections most likely to develop these waves are those in which the canal bottom is raised above theoretical grade. At the point of maximum upward tolerance the depth should be greater than critical depth when computed with the reduced value of *n*.

Coefficient of roughness

Coefficients of roughness (*n*) according to Manning's formula for unlined canals with various characteristics and for the design of different types of linings are given in Table 9 (see pages 60–61).

FIGURE 21. Flap valve installation for a canal underdrain (USBR). All measurements in inches.

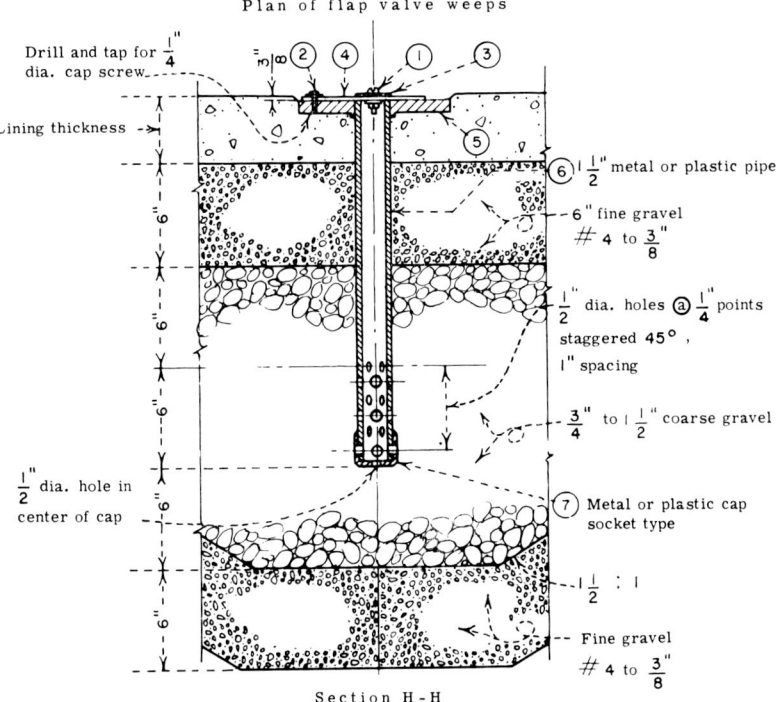

TABLE 9. — MANNING'S COEFFICIENT OF ROUGHNESS (N) FOR UNLINED AND LINED CANALS

Surface conditions	Value of n	Reference
Unlined canals		
Smooth natural earth canals, free from weed growth, little curvature	0.020	(F4)
Small canals in good condition	0.025	(F4)
Earth canals with considerable aquatic weed growth	0.030-0.035	(F4)
Earth canals with thick aquatic weed growth	0.040-0.050	(F4)
Rock canals — main canals	0.030-0.035	France (A16)
— small canals	0.035-0.040	France (A16)
— smooth and uniform .	0.025-0.040	Pakistan (A16)
— jagged and irregular .	0.035-0.050	Pakistan (A16)
Lined canals		
CEMENT CONCRETE		
— exceptionally good finish (rare) . .	0.011	India (A16)
— very well-finished linings	0.013	U.S.S.R. and other countries (A16)
— well-finished for straight canal reaches	0.013	France (A16)
— worldwide adopted value for well-finished linings	0.014	(A1, A16 and others)
— worldwide adopted value for average finished linings	0.015	(A16 and others)
— widely adopted value for poor finish or for curved reaches (France) . . .	0.017	(A16 and others)
— poorly finished, badly maintained canals	0.018	India, Pakistan (A16)
ASPHALTIC CONCRETE		
— machine placed	0.014	United States (A1)
— smooth	0.014	Japan, U.S.S.R. (A16)
— rough	0.017	Japan, U.S.S.R. (A16)

TABLE 9. — MANNING'S COEFFICIENT OF ROUGHNESS (N) FOR UNLINED AND LINED CANALS (*concluded*)

Surface conditions	Value of n	Reference
Lined canals (cont'd)		
SOIL-CEMENT		
— well-finished	0.015	(A3)
— rough	0.016	(A3)
CEMENT MORTAR (hand finished)		
— normal	0.013	Pakistan (A16)
— maximum	0.015	Pakistan (A16)
SHOTCRETE (Gunite)		
— normal	0.017	United States (A1)
	0.018	Australia (A16)
— maximum	0.019	Pakistan (A16)
	0.023	Pakistan (A16)
PRECAST CONCRETE BLOCKS (slabs)	0.015-0.017	Japan (A3)
PRECAST CONCRETE FLUMES	0.012-0.016	Japan (A16)
BRICK		
— brickwork in cement mortar	0.013-0.017	India (A16)
— exposed brick surface (design figure)	0.0146	India (A3)
— exposed brick surface (actual measured value)	0.018	India (A3)
STONEWORK	0.018-0.0225	India (A16)
DRY RUBBLE MASONRY	0.023-0.035	Japan (A16)
EXPOSED PREFABRICATED ASPHALT MATERIALS	0.015	United States (A1)
BURIED MEMBRANE AND COMPACTED EARTH LINING		
— small canals	0.025	United States, Japan (A16)
— large canals	0.020-0.0225	United States (A1)

For easy reference Manning's formula is given below:

$$V = \frac{1}{n} \times R^{2/3} \times s^{1/2} \text{ (metric)}$$

$$V = \frac{1.486}{n} \times R^{2/3} \times s^{1/2} \text{ (English)}$$

where

V = velocity;
n = coefficient of roughness;
R = hydraulic mean radius = A/P;
A = cross-sectional area of water prism;
P = wetted perimeter;
s = longitudinal bed- or water-surface slope.

Prevention of silt deposits in canals

Bedload and fine sand can be prevented from entering a canal, but not suspended silt. In unfavourable conditions suspended silt may settle in the irrigation canal network where its removal would be costly. The minimum velocity which prevents silting is called the critical velocity. The deeper a canal becomes, the higher the velocity necessary to prevent silting. Figure 22 shows critical velocity curves developed by Kennedy for canals in the Punjab (see A15). Curve No. 2 shows that the canal with a 0.0005 slope is safe from silting at flow depths up to 7.25 ft (2.2 m). Curve No. 3 shows that the canal with a 0.0002 slope is safe from silting only to the relatively shallow depth of 3.8 ft (1.2 m).

CEMENT CONCRETE LININGS

Cement concrete linings [1] probably constitute the best type in which benefits justify a high cost. Properly designed, constructed and maintained concrete linings should have an average serviceable life of over 40 years. Some are still in good condition after 60 years. If the dete-

---

[1] In the present text this type of lining is generally referred to as "concrete lining."

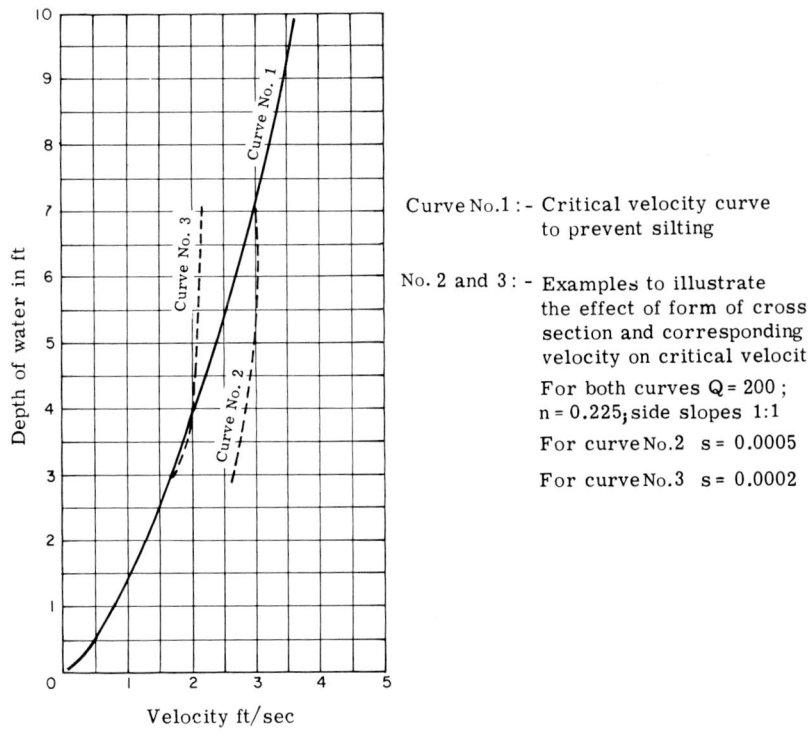

FIGURE 22. Critical velocity curves for the prevention of silt deposits in canals (A15).

riorating action of salts and the development of cracks can be checked or do not occur, the lining may last indefinitely.

Concrete linings are suitable for large and small canals, and for both high and low velocities. They fulfil practically every purpose of lining. They are usually subject to some cracking, but cracks which permit appreciable leakage can be sealed with asphaltic compounds. Costly maintenance is seldom necessary.

Figures 23 through 25 are examples of concrete-lined canals.

Design of cement concrete linings

Thickness of lining. No general rule can be given for establishing the thickness of concrete linings. For small canals and ditches and in locations without severe frost action, unreinforced concrete linings of about 4 cm thickness have been satisfactory.

FIGURE 23. Irrigation canal high on a hillside, Horseshoe Feeder Canal, near Loveland, Colo., U.S.A.

FIGURE 24. Properly lined and maintained canal, Thailand. Note width of embankments.

In most countries with mild climates concrete linings 5 to 8 cm thick for small to medium-size canals and 8 to 10 cm for medium to large canals have been found adequate. Under more severe climatic conditions or in canals with frequent changes in level (e.g., combined irrigation and hydroelectric canals) and on unfavourable subgrades, thickness is increased and may exceed 15 cm for large canals.

Figure 26 shows the thickness of some hard-surface linings in relation to canal capacity as recommended by the U.S. Bureau of Reclamation.

The permissible tolerances are 1 to 1.5 cm (A14).

As a general rule thickness should not be increased to guard against the results of careless workmanship (insufficient curing, poor subgrade, poor concrete, etc.).

Reinforcement. Most concrete linings installed in older irrigation canals were reinforced. During recent years reinforcement has been omitted

FIGURE 25. Concrete lining in a canal on a 10 percent slope, with ladder-type checks for velocity control, Otago, New Zealand.

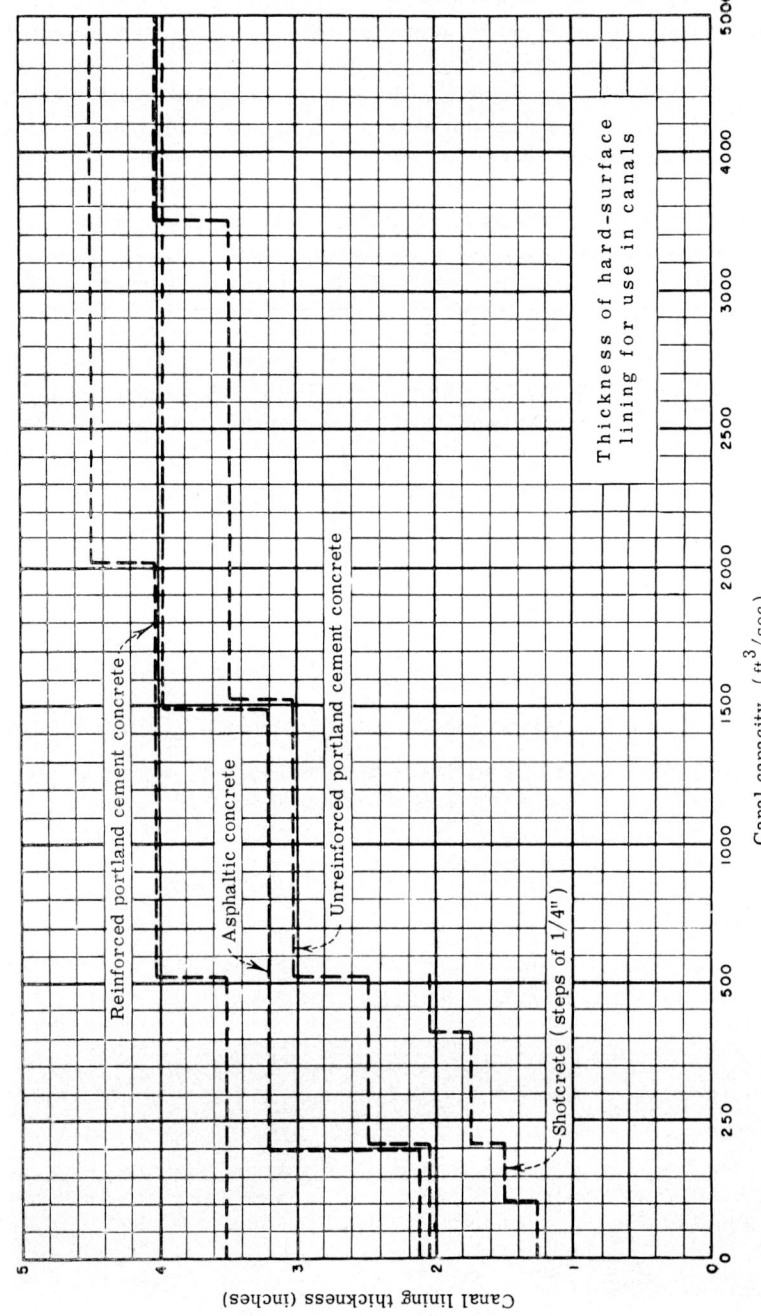

FIGURE 26. Determination of thickness of hard-surface lining based on canal capacity (USBR).

wherever possible to reduce construction costs and because it did not materially improve effectiveness or durability.

Unreinforced concrete linings are more susceptible to damage by hydrostatic or other pressures under the linings than reinforced concrete linings, but not to the degree that the difference in cost might suggest. Where unexpected hydrostatic pressures are encountered under the lining, unreinforced concrete ruptures more readily than reinforced concrete, thus relieving the pressure and reducing the area of damage (A1). The main function of reinforcement is to reduce the width of cracking and prevent separation of the cracked slab.

In the U.S.S.R. prestressed reinforced concrete slabs are used in areas where earthquakes occur and in case of subsiding and swelling soils.

The use of reinforcement steel is of no material benefit if transverse joints are provided at sufficiently frequent intervals to control intermediate cracking. It cannot be justified except under unusual conditions, such as high back pressure, high flow velocities in the canal, unstable subgrade and in reaches where failure would endanger life and property adjacent to the canal.

The necessary area of reinforcement can be found with the following formula:

$$A = \frac{L \times f \times w}{2s}$$

where

- A = the area in square inches of steel per foot of width in the direction in which L is measured;

- L = distance in feet between free transverse joints in computing longitudinal steel or between free longitudinal steel or between free longitudinal joints or edges in figuring transverse steel.

- f = coefficient of friction between slab and subgrade (which varies from 0.5 to 3.0 depending on subgrade material, a value of 1.5 to 2.0 usually being assumed for average conditions);

- w = weight of the concrete slab in pounds per square foot;

- s = allowable working stress in steel in pounds force per square inch (usually assumed to be about one half the ultimate strength of the steel).

The amounts of reinforcement steel generally vary from about 0.10 to 0.40 percent of the area in the longitudinal direction and from about 0.10 to 0.20 percent of the area in the transverse direction (A9).

Joints. Four kinds of joints or grooves are used in concrete canal lining: construction joints, transverse contraction joints, longitudinal contraction joints, and expansion joints.

A construction joint is placed at any location where it is suitable during construction (interruption of work). Usually, it later performs the function of a transverse, longitudinal or expansion joint.

Transverse contraction joints are installed to control transverse cracking which results from shrinkage during volume change caused by drops in temperature or moisture loss, as well as by moisture loss in curing.

The U.S. Bureau of Reclamation recommends the following spacing of joints in unreinforced concrete:

Thickness of lining	*Approximate joint spacing*
5 to 6.5 cm (2 to 2.5 in)	3 m (10 ft)
7.5 to 10 cm (3 to 4 in)	3.5 to 4.5 m (12 to 15 ft)

Average spacing is 50 times the thickness of the slab (A7).

In reinforced concrete linings, joint spacing should be limited to 6 m (appr. 20 ft) to prevent large cracks which may make it difficult to keep joints watertight. It is essential that reinforcement at transverse joints be stopped so that cracks will form at those points.

Contraction joints are usually of the weakened plane type formed by constructing a vertical groove in the top third of the concrete. This groove should be filled with a suitable sealing compound.

Longitudinal joints are used to control irregular longitudinal cracking in unreinforced slabs where the perimeter of the lining is 9 m (about 30 ft) or more and are spaced 2.5 to 4.5 m apart (A1). In reinforced slabs the amount of transverse steel usually used is sufficient to eliminate the need for longitudinal joints except in very large canals.

Expansion joints in concrete canal linings are ordinarily not required, except where the lining abuts on fixed structures or under other extreme conditions. Experience has shown that the use of expansion joints has invariably resulted in increased openings of nearby contraction joints. This is undesirable in canal linings since it increases the difficulty of maintaining watertight joints (A4).

Some common types of joints and grooves are shown in Figure 27. Types *a, b* and *c* have been successfully used in various concrete canal linings. Type *a* is a dummy groove recommended by the U.S. Bureau of Reclamation. Joints *d* and *e* are highly efficient with regard to watertightness. Joint *f* is suitable for thin linings. The sill can be placed ahead and continuous construction is possible if a means of cutting the slab at the joint is provided. Type *g* is suitable not for modern paving machinery but for hand placement. Figure 28 shows a template used to form dummy groove contraction joints.

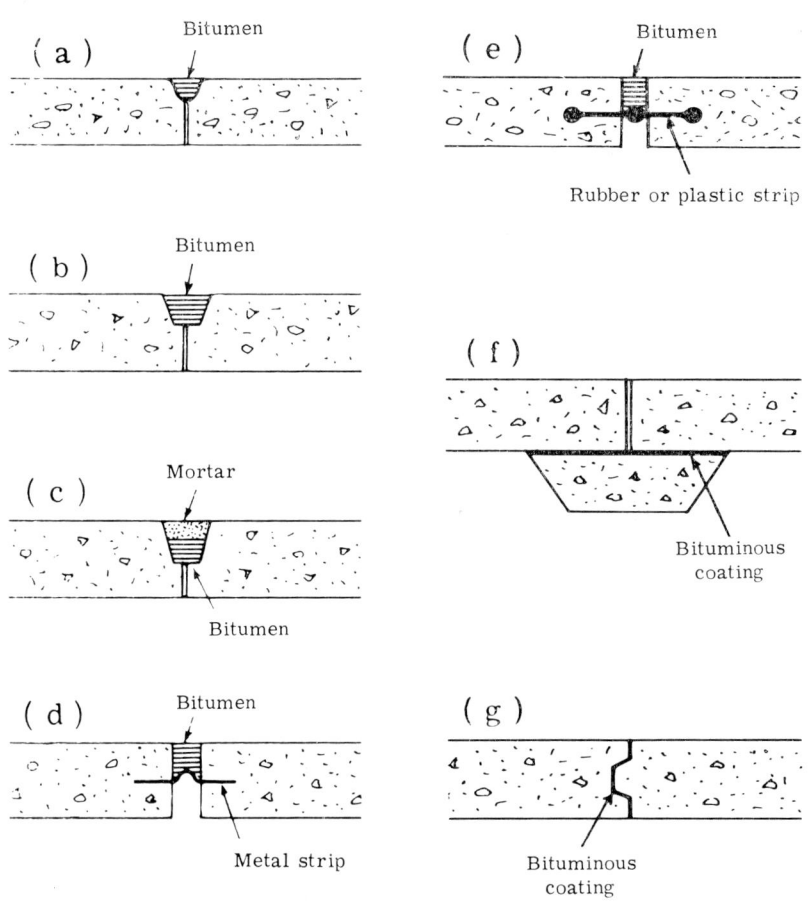

FIGURE 27. Typical joints for concrete canal lining.

FIGURE 28. Forming dummy groove contraction joints with a template.

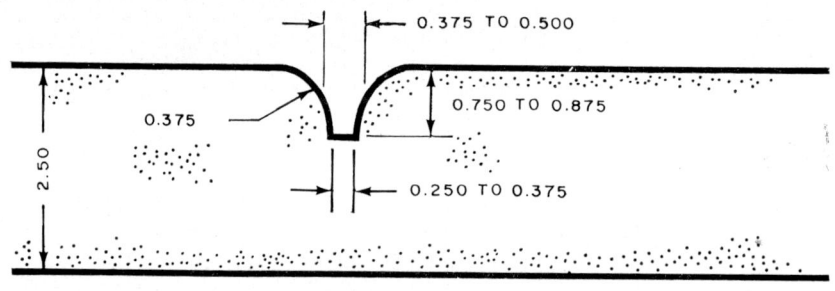

Groove detail

FIGURE 29. Detail of ASAE standard groove (all dimensions in inches).

The ASAE Standard S289, "Concrete slipform canal and ditch linings," prescribes a groove-type joint as shown in Figure 29.

A more detailed discussion of the spacing of grooves and the methods used to form them is found in the U.S. Bureau of Reclamation's *Concrete Manual*.

Standardization of canal sections. To allow standardization of slipforms and other canal construction equipment, standard canal sections have been introduced in several countries. Some of the recommendations worked out by the American Society of Agricultural Engineers are given below.[1]

---

[1] ASAE Standard S289 — "Concrete slipform canal and ditch linings," obtainable from ASAE, 420 Main Street, St. Joseph, Michigan (49085), U.S.A.

TABLE 10. — DIMENSIONS OF STANDARD TRAPEZOIDAL CANAL SECTIONS

Section	Z	a	b	c	e Min	e Max	R
				Inches			
A-1	1:1	14.07	12.00	4.00	15.00	30.00	9.00
A-2	1:1	26.07	24.00	4.00	15.00	30.00	18.00
B-2	1.5:1	25.51	24.00	6.00	24.00	48.00	18.00
B-3	1.5:1	37.51	36.00	6.00	27.00	54.00	18.00
B-4	1.5:1	49.51	48.00	6.00	33.00	66.00	18.00
B-5	1.5:1	61.51	60.00	6.00	36.00	72.00	18.00
B-6	1.5:1	73.51	72.00	6.00	42.00	84.00	18.00

The recommended standards consist of two sections with 1:1 side slopes and five with 1.5:1 side slopes. The dimensional details of these standard sections are shown in Figure 30 and Table 10.

Figure 31 shows the capacity chart for these standard sections. The chart is based on Manning's formula for uniform flow in open channels using an *n*-value of 0.014. It serves as an aid to planners and designers in selecting the most efficient standard trapezoidal canal section. It should be noted that the seven standard sections provide the desirable overlays in carrying capacity which permit adequate flexibility in the choice of a section for optimum economy in materials and other construction costs.

Quality of concrete. Concrete used in canal linings should be mixed so that it is firm enough to stay in place on the side slopes. Since the concrete in canal lining does not act as a structural member, its strength usually is not an important factor. As a general rule, if the concrete is sufficiently durable to resist the wetting, drying, freezing and thawing to which a canal is exposed, it will be strong enough for all but the most extreme conditions. Air-entrained concrete is recommended for all canal work, particularly where exposure to freezing temperatures is anticipated. Air-entrained concrete is also easier to handle and place than non-air-entrained concrete.

Where soluble sulphates, such as sodium, magnesium, calcium or potassium, are present in the soil in appreciable quantities (more than 0.10 percent), either Type II or Type V cement U.S. specifications or equivalent should be used — the latter for extremely severe sulphate conditions (i.e., more than 1 000 ppm sulphate as SO_4). Type II cement is characterized by a relatively low C_3A content. Type V is characterized by low

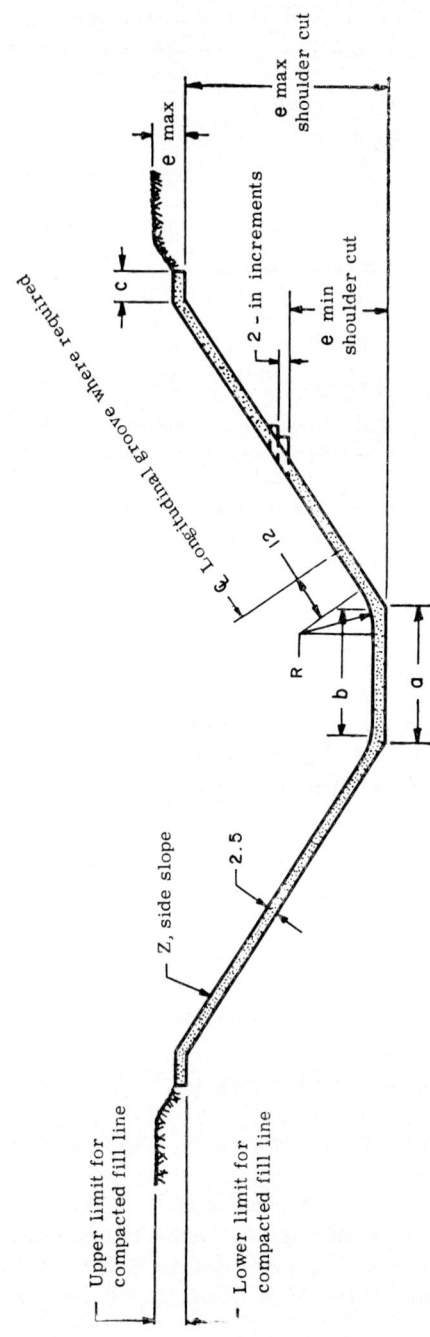

FIGURE 30. Standard trapezoidal canal section (ASAE). All dimensions in inches.

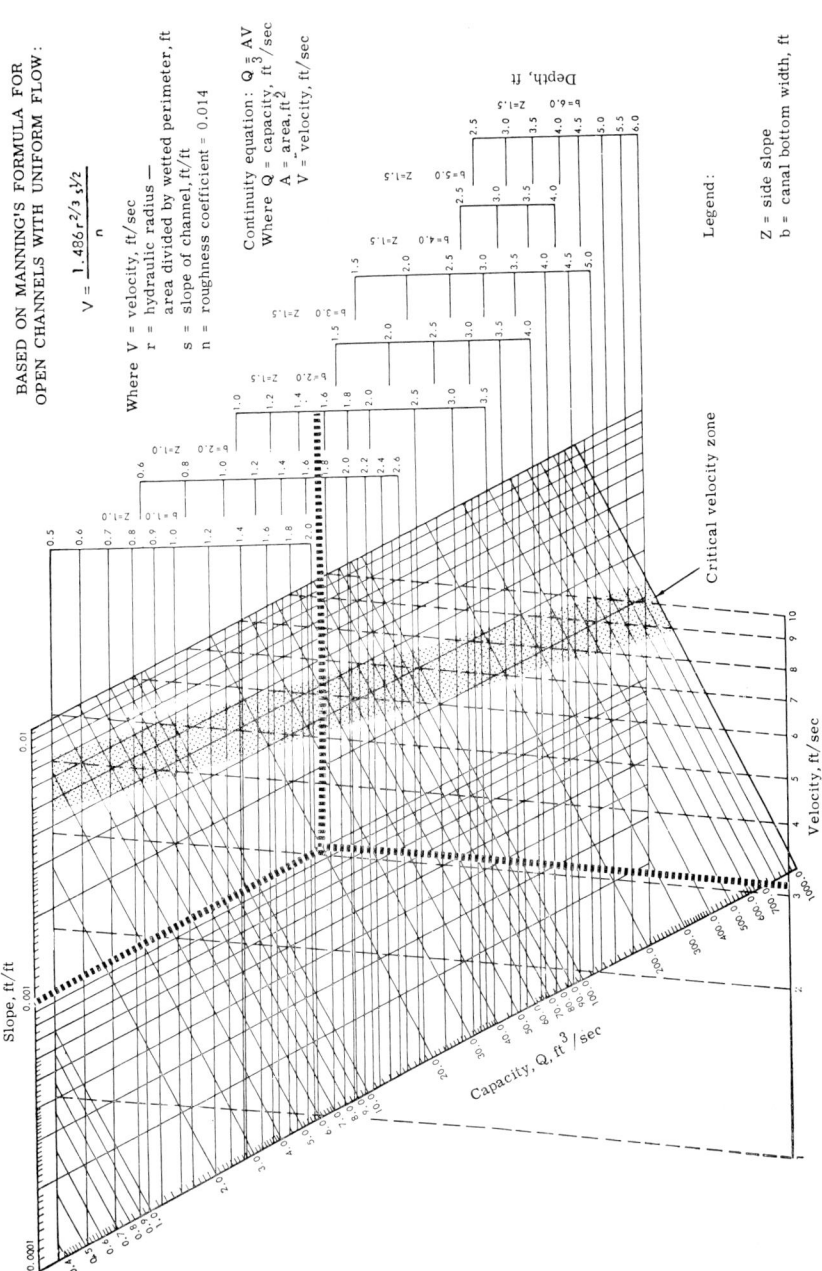

FIGURE 31. Capacity chart for concrete-lined canals (ASAE).

contents of C_3A and C_4AF and by high contents of C_3S and C_2S. (See U.S. Bureau of Reclamation, *Concrete Manual*, F5.) Although these types of cement are preferable, reduction of the sulphate effect can also be obtained through an increase in the cement content.

Figure 32 shows the sulphate-deteriorated wall of a rectangular concrete-lined canal in Mauritius.

The use of hydraulic lime (obtained by burning clayey limestone) in combination with portland cement for canal lining concrete was tested by the U.S. Bureau of Reclamation. The aim of the tests was to develop a lining with less drying shrinkage and able to withstand greater tensile deformation without cracking. Total substitution of portland cement by hydraulic lime failed completely, while a combined use resulted in a decrease in compressive strength of the concrete in 28 days, in proportion to the amount of hydraulic lime used. Controlled shrinkage tests of concrete containing 10 and 20 percent hydraulic lime indicated less resistance to cracking than the control specimens, in which the binding agent was 100 percent portland cement.

The U.S. Bureau of Reclamation also investigated dry-tamped concrete because the lower water content might result in fewer cracks from dry shrinkage and some economy might be realized from the lower cement content.

Dry-tamped or zero-slump concrete of 0.44 water-cement ratio exhibited considerably less drying shrinkage, greater durability, and about the same

FIGURE 32. Concrete lining deteriorated by aggressive water during 14 years of service, Mauritius.

strength and permeability as 9 cm slump concrete of the same water-cement ratio. Furthermore, the dry-tamped concrete contained ca. 60 to 90 kg of cement less per m³ than the 8 cm slump concrete with which it was compared. It was concluded that, until labour-saving methods for placing dry-tamped concrete are developed, the high cost of hand placement would far overshadow any benefits from these better characteristics or the saving in cement (A1). However, where labour is available at low cost and where cement is in short supply, the use of dry-tamped concrete apparently offers economic advantages.

Design example. Design a concrete-lined canal from the following data:

Full supply capacity = Q = 1 800 ft³/sec = 50.9 m³/sec;

Longitudinal bed slope = S = 0.0001;

Side slopes = 1.5:1;

Coefficient of roughness (Manning's n) = 0.015;

Ratio of bed width (B) to water depth (y) should be about 1 for small canals and about 3 for large canals.

This canal is in the middle range; therefore try ratio 2.

Using Manning's formula (see pages 58-62), with B = 23 ft = 7 m:

y = 11.81 ft (3.70 m);
A = 480.84 ft² (44.7 m²);
V = 3.74 ft/sec (1.14 m/sec);
R = 7.33 ft (2.24 m).

Check: (1) B/y = 1.95 — OK.

Check: (2) v = 3.74 ft/sec, which is less than 8 ft/sec (2.5 m/sec), the maximum recommended velocity in unreinforced concrete linings (see page 55) — OK.

Check: (3) critical depth: using n = 0.012 gives y = 10.6 ft (3.23 m), which is greater than critical depth of 5.1 ft (1.56 m) — OK.

Lining freeboard (from Figure 16 on page 53) = 1.9 ft (0.58 m).

Bank freeboard (from Figure 16 on page 53) = 4.0 ft (1.22 m).

Construction of cement concrete linings

Subgrade preparation. Concrete linings should be placed only on a well-consolidated subgrade. Fills or embankments should have been compacted by rolling, tamping or vibrating. For small canals, satisfactory compaction may be obtained by ponding water in the canal before final trimming of the subgrade. On embankments, where rollers, tampers or vibrating equipment are used, the material should be placed and compacted in approximately 15 cm layers at a predetermined optimum moisture content.

After preparation of the subgrade, care should be taken to prevent any great loss of moisture prior to placement of the concrete lining. In all cases, just before the concrete is placed, the subgrade should be sprinkled in a manner which will not form mud or puddles.

With the use of subgrade-guided slipforms, alignment and grade of the finished lining depend almost entirely on the care and accuracy with which the subgrade is prepared. According to ASAE Standard S289 "Concrete slipform canal and ditch linings" (adopted June, 1965) the following tolerances for lining thickness, alignment, and grade are considered adequate to assure quality installations:

Item	*Tolerance*
Departure from established alignment	5 cm on tangents 10 cm on curves
Departure from established profile grade	2.5 cm
Reduction in lining thickness	10 percent of specified thickness

Available excavating and lining equipment will permit construction of linings within the above tolerances.

Concrete mixes. As stated before, concrete for lining a canal should be plastic enough to consolidate thoroughly and stiff enough to stay in place on the side slopes.

For the lining of farm irrigation ditches with concrete, the following proportions have been suggested (C3):

1. The slump should not exceed 7.6 cm (3 in) for hand placing or 12.5 cm (5 in) for placing with the slipform.
2. In mild climates a minimum of 270 kg/m^3 of concrete (4½ sacks per yd^3) with a water-cement ratio of not more than 0.70 should be used.
3. In moderate climates a minimum of 330 kg/m^3 (5½ sacks per yd^3) with a water-cement ratio of not more than 0.60 should be used.

Concrete aggregate should be clean, hard and durable. The maximum size of coarse aggregate should not be larger than one half the thickness of the lining.

The U.S. Bureau of Reclamation accepts the following maximum permissible water-cement ratios for canal linings:

Mild climate, rainy or arid, rare snow or frost	0.58 ± 0.02
Severe climate, wide range of temperature, frequent freezing and thawing	0.53 ± 0.02

For small operations, where time and personnel are not available to determine the proportion of the concrete mix according to established procedures, the approximate mixes may be chosen from Table 11 (derived from F7):

TABLE 11. — APPROXIMATE CEMENT CONCRETE MIXES

Maximum size of aggregate	Approximate cement content per cubic metre of concrete in kilos	Kilos of aggregate per cubic metre of concrete		
		Sand (dry)		Gravel or crushed stone
		Air-entrained concrete	Concrete without air	
1.5 cm	385	850	890	930
2.5 cm	365	790	825	1 010
5 cm	330	710	750	1 200

In Uzbekistan (U.S.S.R.) a maximum ratio of water to cement of 0.60 to 0.65 has been employed for concrete linings. The cement expenditure ranged within the limits of 250 to 300 kg per cubic metre (B29).

ASAE Standard S289 requirements for slipform concrete linings are shown in Table 12 on page 78.

The aggregate should conform to ASTM C33 standards. With aggregate materials of other qualities a minimum cement content of 385 kg/m³ (6.5 sacks/yd³) should be used.

Usually a mix with an aggregate of sand is needed for machine-placed lining to give adequate workability. Consistency of concrete is important because a variation of as little as 2.5 cm (1 inch) in slump can seriously interfere with the progress and quality of work.

Finishing and curing. Concrete specifications should always include a paragraph on curing. Proper curing greatly improves the durability, wear-resistance and watertightness of concrete. Tests have shown that

TABLE 12. — ASAE SPECIFICATIONS FOR SLIPFORM CONCRETE

Class of concrete	Compressive strength at 28 days	Maximum size of aggregate	Cement content
Normal concrete			
Mild exposure	210 kg/cm^2 (3 000 psi)	appr. 20 mm (0.75 in)	325 kg/m^3 (5.5 sacks/yd^3)
Average frost exposure	245 kg/cm^2 (3 500 psi)	appr. 20 mm (0.75 in)	355 kg/m^3 (6 sacks/yd^3)
Air-entrained concrete			
Mild to average frost exposure	210 kg/cm^2 (3 000 psi)	appr. 20 mm (0.75 in)	340 kg/m^3 (5.75 sacks/yd^3)
Severe frost exposure	210 kg/cm^2 (3 000 psi)	appr. 20 mm (0.75 in)	385 kg/m^3 (6.5 sacks/yd^3)

concrete which was moist-cured for 14 days had a 28-day strength about twice that of concrete which was allowed to dry in the air. Cooke (B4) states that the omission of curing is about as sensible and has about the same effect as leaving out one third to one half of the cement.

A smooth, hand-trowelled surface finish would increase the carrying capacity of the canal and would be justified where labour is inexpensive. However, where growth of moss or algae and silting is likely to occur, the additional effort may not be warranted and a reasonably smooth surface without voids will be adequate. The concrete must be kept continuously saturated to provide adequate curing — that is, the exposed surface must be kept moist for three to five days, or the water part of the mix must be sealed in with a sprayed-on curing compound or other impervious membrane.

ASAE Standard S289 prescribes the application of a pigmented compound within 20 minutes after placing the concrete. Coverage should not exceed 5 m^2 per litre.

Figure 33 shows the curing of a cement concrete canal at Digod Farm, Chambal Project, India, by ponding or by using wetted jute sacks. Curing generally lasted 14–21 days depending on seasonal and climatic conditions.

Manually placed concrete linings. Placing concrete linings by hand may prove economical when low-cost labour is available or when the reach of canal is too short or the cross section too small to be economical

FIGURE 33. Curing of a cement concrete lining, Chambal Project, India.

for mechanized placing. A hand method for lining small irrigation ditches with parabolic cross sections, described by Zimmerman (A15), is explained in the following paragraphs.

The canal lining is built between two 5 × 10 cm (2 × 4 in) parallel boards which are placed upright on top of the embankment. These boards are then utilized as guides for all subsequent operations — repair of earthwork, trimming, fastening of the screed guides, formwork, etc. (Figures 34–37). As all operations are based on these guide boards, they must be carefully aligned, levelled, and fastened securely. The screed guides can be made of T-profile steel bent to the shape of the canal cross section and fastened to the guide boards.

Concreting is done in alternate sections (Figure 35) between a set of two screed guides by shovelling the concrete against the sides of the excavation (Figure 36). The guide boards serve as shuttering along the top of the canal and the screed guides serve as transverse shuttering. The concrete is then compacted, formed and smoothed by using a screed (Figure 37). This is dragged two to three times on top and along the guides. After the concrete has set sufficiently, the alternate sections are filled in the same way. This time the concrete of the previously cast adjacent sections serves as a guide (Figure 38), and the screed guides are moved on to another pour.

For a trapezoidal canal, where the side slopes are straight, a non-reinforced concrete lining 3.5 cm thick is usually ample since the sides

FIGURE 34. Parallel boards and screed guides for hand-placed lining.

FIGURE 35. Placing alternate sections between screed guides.

FIGURE 36. Concrete for lining being shovelled against subgrade.

FIGURE 37. Concrete being compacted, formed and smoothed with a screed.

FIGURE 38. Concreting between set sections, using them as guides.

can be easily and well compacted with a hand tamper. This allows the use of concrete with a very low water-cement ratio and low slump.

For a parabolic canal a concrete lining 5 cm thick has proved best, as here the curved surface with the steep side slope does not facilitate the use of a hand tamper and only screeds can be used for compaction.

By pouring alternate sections at intervals of about a week, most of the damage due to drying shrinkage is prevented and the construction joints between the older and newer sections serve as expansion joints.

The freeboard recommended for this type of canal lining should be 7.5 cm for a canal without structures which influence the flow and 10 cm for a canal with turn-outs and checks. Where water control and canal maintenance are unreliable, the freeboard must be higher.

In the manner described it is possible to construct a section of about 75 to 90 m (250 to 300 ft) of concrete lining with a crew of one supervisor and two semiskilled and 13 unskilled labourers in an 8-hour day, excluding excavation.

Figures 39–41 show a steel form for small rectangular canal sections. These forms have been successfully used in Iran. Steel was used because of the lack of wood, hardboard or chipboard. One form was used 150 times without showing significant wear. The inside surfaces of the forms must be well cleaned and oiled between uses.

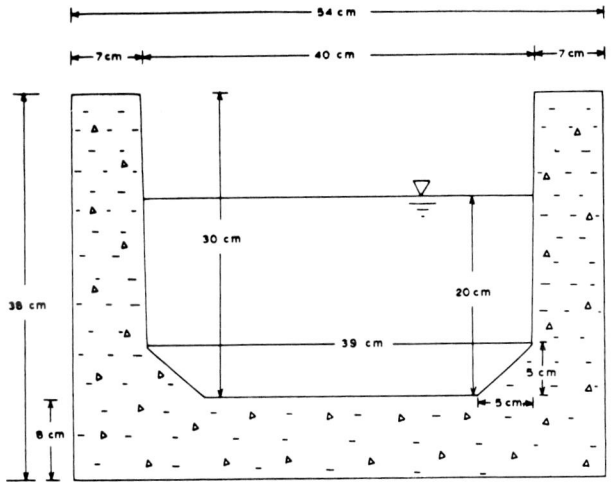

Section through concrete lining

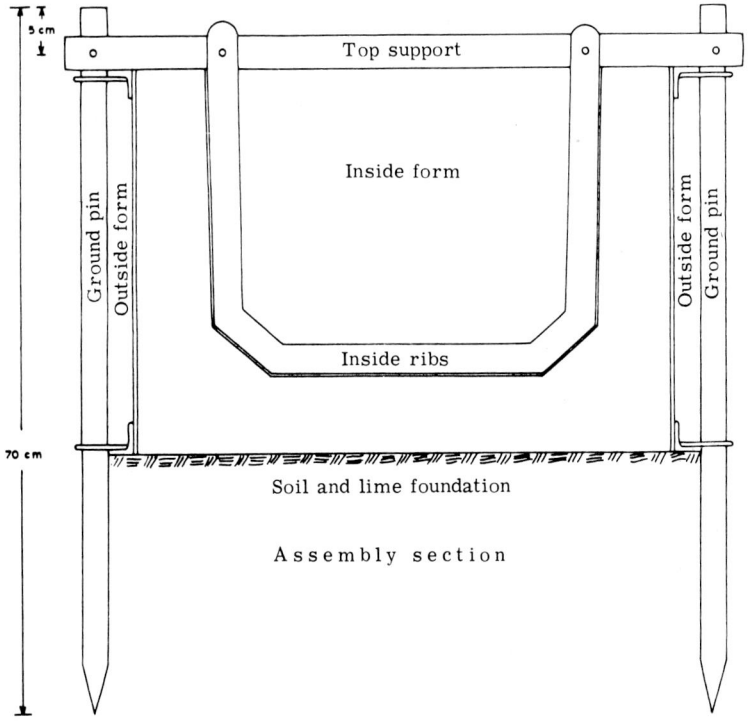

Assembly section

FIGURE 39. Steel concrete form for rectangular canal sections.

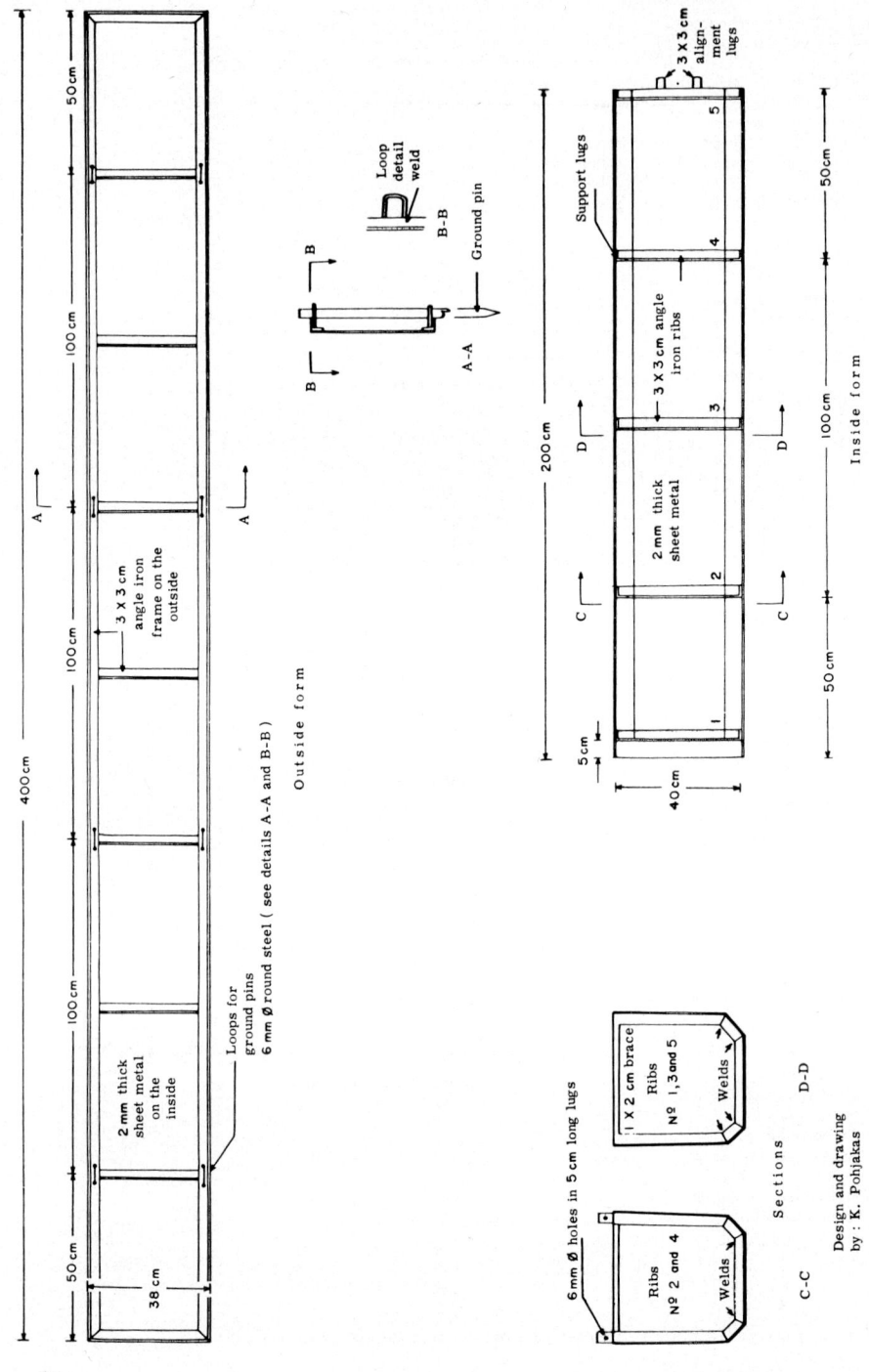

84

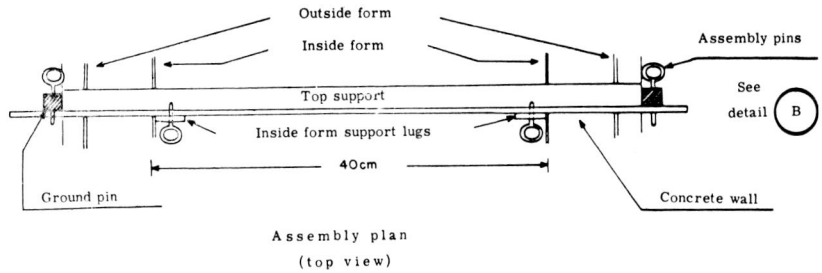

Assembly plan
(top view)

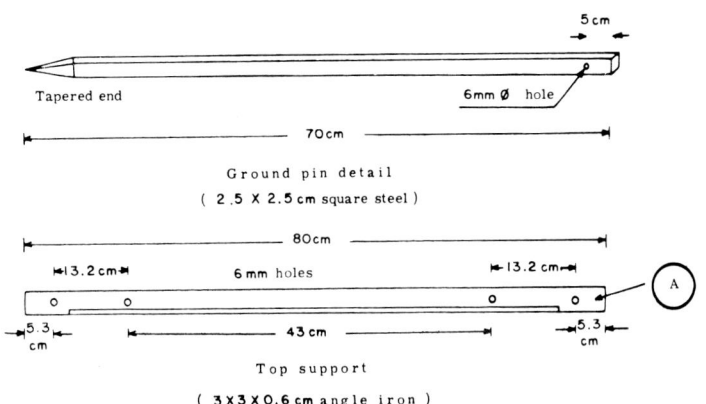

Ground pin detail
(2.5 X 2.5 cm square steel)

Top support
(3 X 3 X 0.6 cm angle iron)

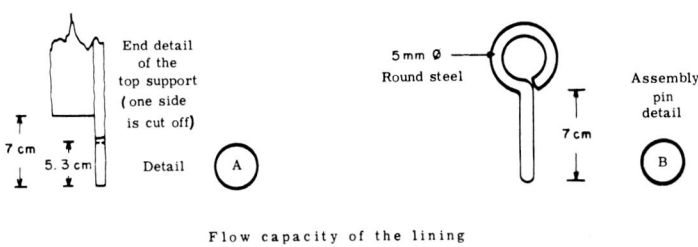

Flow capacity of the lining

1. placed on 0.05 % slope = 24 lt/sec
2. placed on 0.1 % slope = 34 lt/sec
3. placed on 0.2 % slope = 49 lt/sec

FIGURES 40—41 *(left and above).* Steel concrete form for rectangular canal sections.

FIGURE 42. Detail of template used for panel-formed concrete linings.

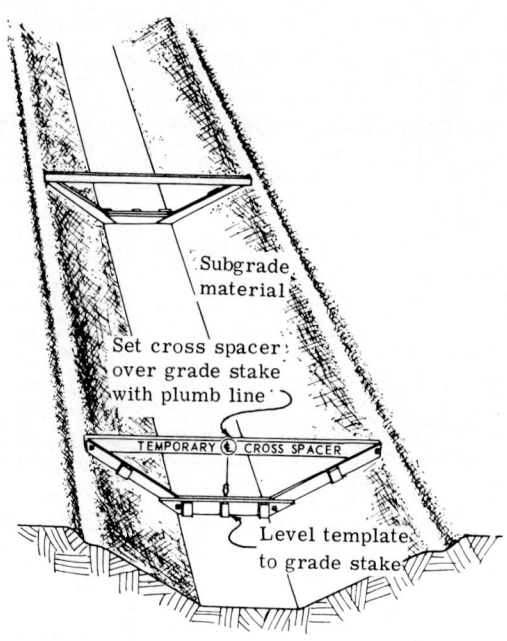

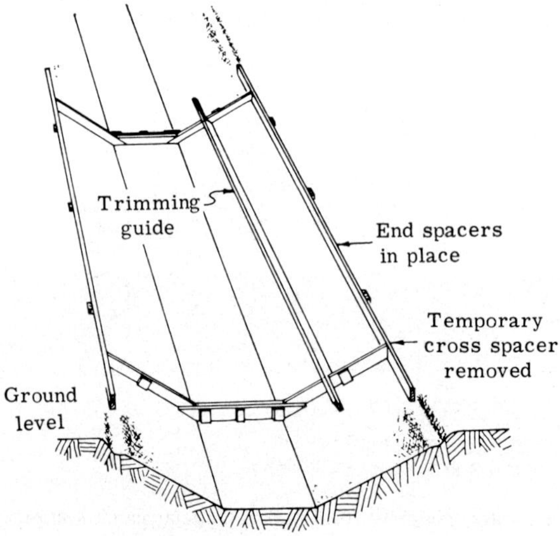

FIGURE 43. Panel form in place and ready for concrete to be poured.

A hand-mixed concrete of 3.5- to 4-in slump is used and worked in by sticks. The capacity is about 10 m per man-day; it is estimated that this could be doubled by using a small power mixer.

For larger ditches and small canals Lauritzen (C58) describes the panel method of hand placing concrete as follows:

After the ditch has been excavated, grade stakes are set along the centre line at intervals similar to the length of the panel to be used and at elevations corresponding to the top of the finished lining.

The templates are usually 5 × 10 cm (2 × 4 in) on edge (Figure 42). The bottom member of the template is levelled with the top of the grade stake and the cross spacer bolted in place temporarily. The cross spacer is then centred over the bottom piece by means of a level or plumb bob so that both sides will have the same slope. The form is then fastened in place by driving stakes into the subgrade along both sides and nailing the form to them.

The second form is installed in the same manner at the next grade stake, 3 m (10 ft) from the first if the panels are to be 3 m long. A 5 × 10 cm piece, the end spacer, is then staked parallel to the ditch along the top between the two cross members (Figure 43). The end spacer provides the top of the form. The temporary cross spacers are then removed from the cross members and the forms are in place.

As an aid to fine grading, a 5 × 12 cm piece, slightly longer than the panel, is notched to a depth equal to the thickness of the concrete lining and fitted over the end forms. This piece is used as a guide in trimming the subgrade to the specified depth.

For a panel-formed lining, a relatively stiff concrete mix — a mix with a slump of 5 to 7.5 cm — is required. The freshly mixed concrete is placed in the formed area and spread with a shovel.

The bottom is poured first, and then the fresh concrete screeded up the slope. If a rope is placed on the screed board, a worker standing at the top of the slope can pull the screed up the slope and assist those below. As the screed board is sawed back and forth across the forms, concrete is added with a shovel ahead of the board to fill minor depressions and to keep a little excess material ahead of the board.

A wooden float can be used to fill in small depressions, such as those caused by stones pulled under the screed, and to touch up the lining, if needed. If screeding is done properly, little finishing with the float is required.

The lining is constructed in alternate sections as previously described. In constructing the connecting sections, the finished sections are used as end forms. The end spacer is staked against the top of the finished sections to complete the forming (Figure 44).

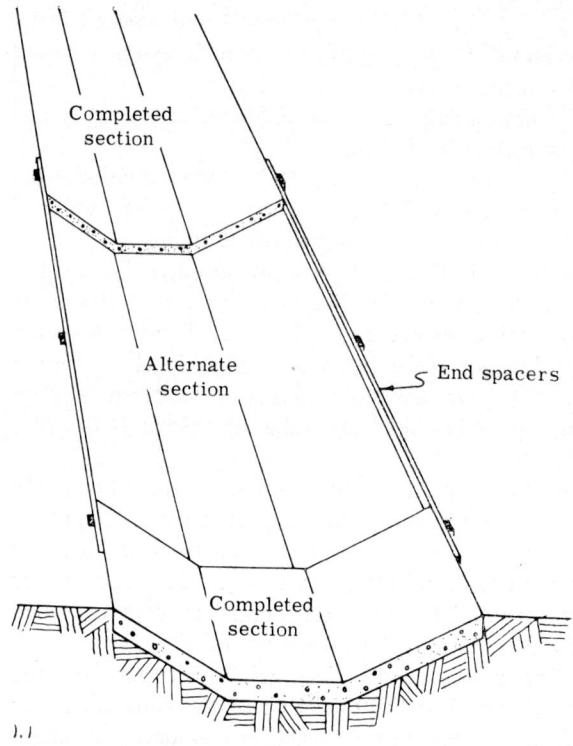

FIGURE 44. End spacers in place on alternate sections of panel-formed lining.

Figure 45 shows the manual construction of a 5 cm concrete lining for a small canal.

Another example of manually constructed concrete lining is the San Francisco River Project, Brazil, executed in 1959 (Figure 46). The canal, which has a design capacity of 2.70 m^3/sec, is lined with 12 cm thick concrete. The joints are of the type shown in Figure 27f. Canal dimensions are: 1.5 m bed width; 1.15 m maximum water depth; 1.5:1 side slopes. The concrete was taken by wheelbarrow to the canal, placed by hand and mechanically vibrated. The surface was hand finished with a 1:3 mortar and cured by a cover of moist soil.

The following paragraphs give two other examples of time and labour requirements for manually installed concrete linings:

Twenty labourers were used to mix and place concrete in alternate sections of a canal in Queensland, Australia (A3). A second group

FIGURE 45. Concrete lining of a small canal by hand.

(a) Trimming canal section.

(b) Placing the 5 cm lining.

(c) Hand screeding the concrete.

(d) Final finishing of the lining.

FIGURE 46. Concrete lining of a canal on the San Francisco River Project, Brazil.

(*a*) Placement of transverse joints.

(*b*) Subgrade with finished joints.

(*c*) Placing 12 cm concrete lining.

(*d*) Completed canal reach.

followed and completed the gaps. The canal was constructed with a bed width of 1.20 m, side slopes of 1.5:1 and a concrete of 5 cm thickness. The concrete was mixed in a 0.4 m^3 (0.5 yd^3) mixer set up at about 45 m intervals and leapfrogged as the work progressed. The concrete, having a slump of only 1.3 cm, was wheelbarrowed to the canal and compacted with vibrating floats (Figure 47). This procedure produced an average of 1 000 m^2 (1 200 yd^2) per shift.

On the Nzoia River Pilot Irrigation Scheme, Kenya, a reach of a canal in highly permeable soil was lined with a 7.5 cm concrete lining placed on an appr. 15 cm grouted rubble foundation (see Figure 48). A 3 m section with a bottom width of 1.30 m, 1.20 m depth and 1.5:1 side slopes required 3.4 m^3 (120 ft^3) of grouted rubble foundation and 1.7 m^3 (60 ft^3) of a 1:2:4 mix concrete. The concrete was produced by mixers and transported with wheelbarrows. Two masons and eight general labourers required a total of 15 ten-hour working days to complete a 46 m (150 ft) reach of this canal.

Slipform concrete linings. For many years concrete canal linings have been built with longitudinally operated lining equipment, commonly called slipforms. After excavation and trimming of the subgrade, pouring, shaping, compacting and smoothing of the concrete lining are done with the slipform. Slipforms are used for almost every size of irriga-

FIGURE 47. Vibrating screed and small float (A3).

FIGURE 48. Hand-placed concrete lining of a canal in the Nzoia River Pilot Irrigation Scheme, Kenya.

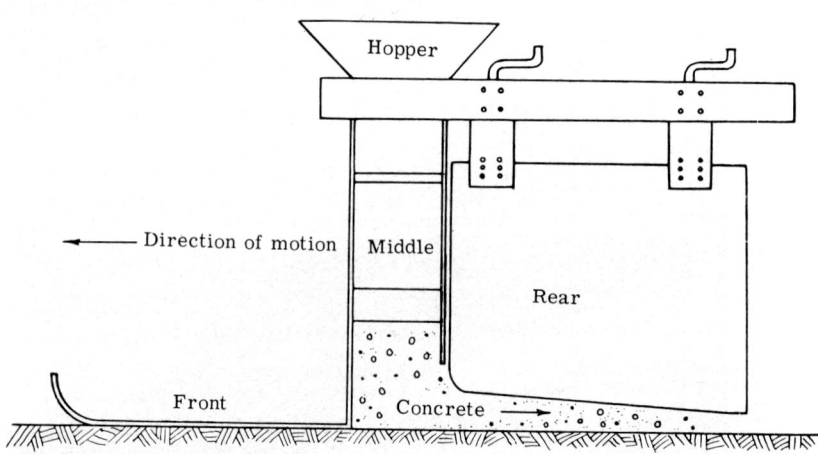

FIGURE 49. Drawing of a type of slipform used to build concrete linings (C58).

tion canal. For small canals they are usually subgrade-guided, and for large canals, with lining perimeters exceeding 7.5 m (25 ft), they are usually supported on wheels, crawlers or rails.

A sectional drawing of the principle of the subgrade-guided slipform is shown in Figure 49. The front part of the form rides on and is guided by the subgrade as it is pulled forward. A winch mounted on the slipform with the cable attached to a deadman or a tractor with a cable attached to the slipform may be used to pull it. The middle section of the slipform is the hopper through which the freshly mixed concrete is poured and distributed to the sides and bottom of the invert. The rear section of the slipform is the strike-off, or screeding, mechanism. The difference in height between the bottom of the rear section and the bottom of the front section determines the thickness of the concrete lining. Since the slipform is guided by the subgrade, the lining will have the same grade and alignment as the invert. Although slipforms vary, two fairly satisfactory designs are illustrated in Figures 50 and 51.

In their booklet *Plans for concrete slipforms* (C61), Lauritzen and Griffin have published design drawings of two bottom-guided slipforms, one for 30 cm (1 ft) and another for 60 cm (2 ft) bottom width. Figures 52 and 53 show the slipform from front and rear. The slipform may be

FIGURE 50. A subgrade-guided slipform (C58).

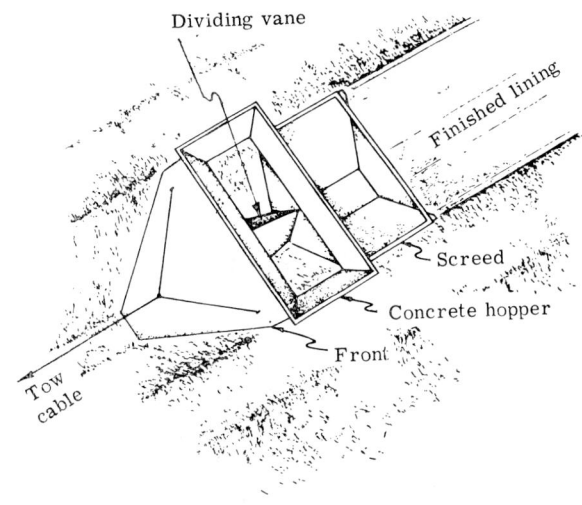

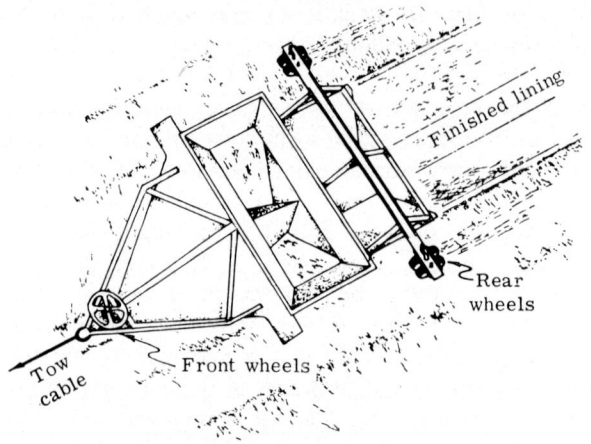

FIGURE 51. Slipform shown in Figure 50 with wheels added to convert it into a berm-guided slipform (C58).

FIGURE 52. Front view of a bottom-guided slipform for standard concrete lining.

equipped with a winch and cable for self-towing as shown in Figure 52 or towed by a tractor as shown in Figure 53.

The 30 cm form can be used in canals with a minimum depth of 50 cm and a maximum depth of 75 cm. The 60 cm form with 1:1 side slopes has the same minimum and maximum depths. The 60 cm form with side slopes of 1.5:1 can be used in canals with a minimum depth of 60 cm and a maximum of 120 cm.

Concrete should be delivered in sufficient volume for the slipform to operate continuously. Usually the concrete is supplied to the form by ready-mix trucks.

During the construction of concrete linings with slipforms no account is taken of turn-outs or any other small structures to be built along the canal. It is cheaper and more convenient to build the canal in its entirety and to break out any section later wherever necessary.

A plough-type ditcher is most commonly used to excavate the channel. In order to obtain the best results when excavating with the plough-type ditcher, several shallow cuts should be made with successive passes rather than deep cuts with fewer passes.

FIGURE 53. Rear view of a bottom-guided slipform for standard concrete lining.

FIGURE 54. Subgrade-guided slipform from front.

FIGURE 55. Subgrade-guided slipform from rear.

The slipform shown in Figures 54 and 55 was used for lining a canal of 90 cm bed width and 75 cm depth in Queensland, Australia (A3). The form is propelled by a hand winch mounted on it and the concrete is transported to the hopper by wheelbarrows. Vibration is applied as the concrete passes from the hopper under the form.

With a water-cement ratio of 0.65 and a slump of 5 cm the form itself gave an excellent finish. With a water-cement ratio of 0.53 and no slump, some tearing occurred, requiring much hand trowelling of rough patches. Concrete was applied at a rate of about 1 m^3 per 5 m (1 yd^3 per 13 ft) of canal. The rate of placement varied from 0.60 to 1.20 m (2 to 4 ft) per minute. Under Queensland conditions it was felt, as of 1957, that the construction of a slipform was warranted if it could be used over 6 km (4 mi) of canal. Placing of slipform lining in small farm ditches at a rate of 1.6 km per day has been reported from Arizona (A15).

Figure 56 shows the lining of a 5 cm thick concrete canal of 90 cm bottom width, 1.25:1 side slope and 135 cm depth, using a slipform with

FIGURE 56. Subgrade-guided slipform equipped with screw hopper.

FIGURE 57. Canal slipform paver capable of paving various depths.

screw hopper. Concrete was placed at a rate of 230 m³ per eight-hour shift.

A plough-type lining machine can be used for small canals with a maximum bottom width of 60 cm and a depth of 86 cm. The machine combines plough, trimmer and liner in one unit. Normally the machine operates by making several passes with the plough, to excavate and compact the soil. Three-inch-slump concrete is dumped into the slipform hopper from transit mixers.

The slipform paver shown in Figure 57 is designed for medium-size canals of different depths. The maximum operating speed is about 3 m per minute, the capacity ranging from 60 to 120 m³ concrete per hour.

A method of constructing a near-semicircular canal of various sizes has been developed by Barragan in Spain (C36). By the end of 1975 about 400 km of canal in Spain were lined by this method.

The canal surface is an arc described by an angle of 160° to 130° (sexagesimals). Diameters range from 1.5 to 10 metres. A great advantage is that relatively dry concrete can be used. The lining consists of a low-quality concrete base with a higher quality concrete surface layer.

This circular section is more suitable hydraulically as well as more stable than trapezoidal or rectangular sections and more economical.

Figures 58 and 59 show the construction of the Orellana-del-Plan-Badajoz Canal having a design capacity of 15 m³/sec (530 cusecs). A

FIGURE 58. Lining a semicircular canal in Spain.

FIGURE 59. Semicircular concrete-lined canal in Spain.

FIGURE 60. Construction of the main canal, Columbia Basin Project, Washington, showing (*in order from foreground*) subgrade trimmer, slip-form trimmer, jumbo for cutting dummy groove contraction joints, and jumbo for applying membrane-curing compound.

36-km long reach of this canal was concrete lined in 10 months (which included a very wet winter) using four sets of excavation, trimming and lining machines.

The lining of a large canal with rail-supported slipforms is shown in Figure 60. The planning, design and construction features of such large canal projects are too individual to be discussed in this publication.

Semimechanized construction method. Where lengths of large canals are too short to warrant the expense of using slipforms, or where adequate equipment is not available and labour is cheap, winch-drawn screeds operating transversely on the canal slopes (see Figure 61) are usually found to be economical. Normally the lining is built in alternate panels and the construction joints between panels then function as expansion joints.

Figures 62 shows a sketch of a sliding screed used on a project in Queensland (A3). The form was moved by two hand winches set at the top of the slope. The concrete was fed to the working platform from the top of the slope by means of a chute. Speed of travel was about 90 cm (3 ft) per minute.

FIGURE 61. Placing concrete on the slope of Pole Hill Canal, Colorado (U.S.A.).

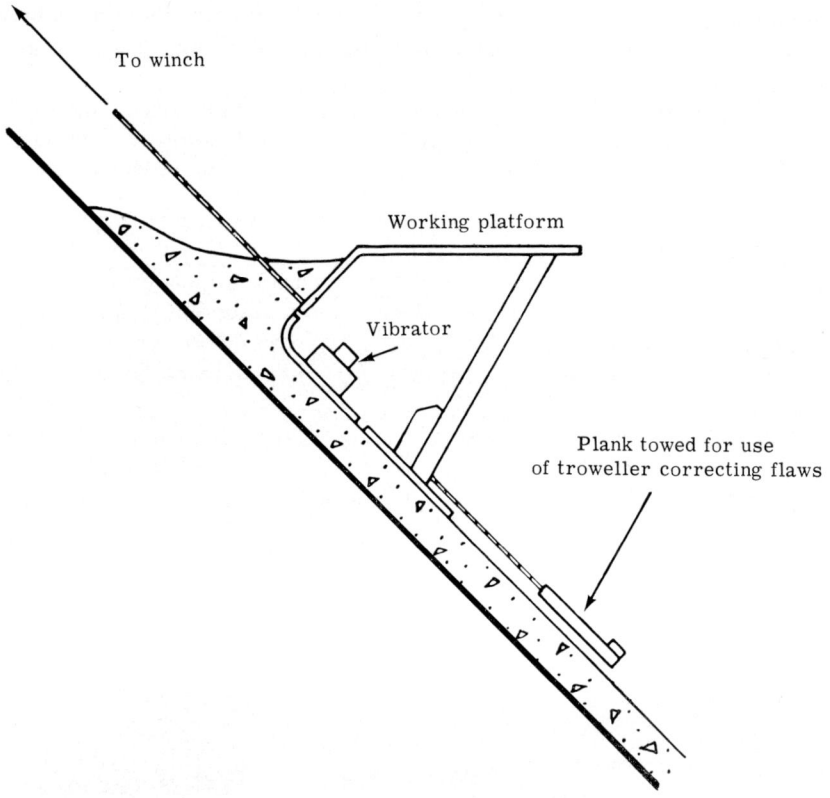

FIGURE 62. Sliding screed for operating transversely up the slope of a canal (A3).

The main features in the design of this form are that the vibration must be applied under a head of concrete and the float section must be reasonably free from vibration. The form must be weighted in order to keep it in solid contact with the screed guides so as to control the thickness of the lining.

PNEUMATICALLY APPLIED MORTAR

These linings are constructed by shooting a mortar consisting of a thorough mixture of cement, sand and water into place by means of compressed air. The usual mix is one part cement to 4.5 parts of sand and a little less water than the amount that would cause sloughing.

The thickness of the linings varies from about 2.5 to 7.5 cm (1–3 in) and wire-mesh reinforcing is sometimes used.

The advantages of this method of construction are the following:

— equipment units are relatively small and easily moved;

— fine trimming of subgrades, precise alignment and close conformity with design dimensions are not commonly required.

Since coarse aggregates are not included in the mix, this method requires greater proportions of cement than concrete lining.

The rate of placement is slow in comparison with slipform concreting and skilled labour is required to operate the equipment and to control the lining thickness.

Pneumatically applied mortar is not generally economical for large lining projects; however, for short canals, for parts of canals with frequent sharp curves, for curve sections of precast concrete linings, for canal reaches in weathered rock and for the repair of canals, it has given excellent results. Construction details are found in A15.

GROUTED FABRIC MATS

In recent years synthetic fabric formworks (mats) pneumatically filled with concrete grout have been used quite successfully for erosion protection and seepage prevention. The main advantage of this lining is that it can be installed in operating canals without dewatering. However, because of its relatively high cost, it has a very limited economical application in irrigation canal lining.

PRECAST CONCRETE SLAB OR BLOCK LININGS

Precast concrete has been used in lining works throughout the world for all sizes of canals, but the trend is declining in countries with expensive labour. This discussion refers only to precast concrete slabs and small trough-formed precast sections for farm channels; precast elements for flumes and pipes are discussed in Chapter 3.

Precast slabs or blocks are usually made 5 to 7 cm thick. Widths and lengths are varied to suit canal dimensions and to provide weights that can be conveniently handled by one or two workers. The following sizes have been used extensively (all in centimetres): India-Maharashtra, $61 \times 61 \times 5$; Morocco, $50 \times 25 \times 7$; Pakistan, $38 \times 38 \times 5$; Portugal, $50 \times 20 \times 6$; U.S.A., $61 \times 20 \times 5$.

In the designs shown in Figure 63, tongue and groove joints are provided along the edges. Other shapes of edges have proved successful and even plain joints are often used. The joints should be sealed with mortar or bituminous mastic. Contraction joints at 4 to 5 m spacing are required if the slabs are joined with cement mortar. For suitable sealing of contraction joints, see page 68.

In some projects, joints have been shaped to accommodate steel bars. However, considering the high cost involved, reinforcement is generally not warranted. Instead, a certain allowance should be made for repair of the damages which may occur from uneven subgrade settlement, etc.

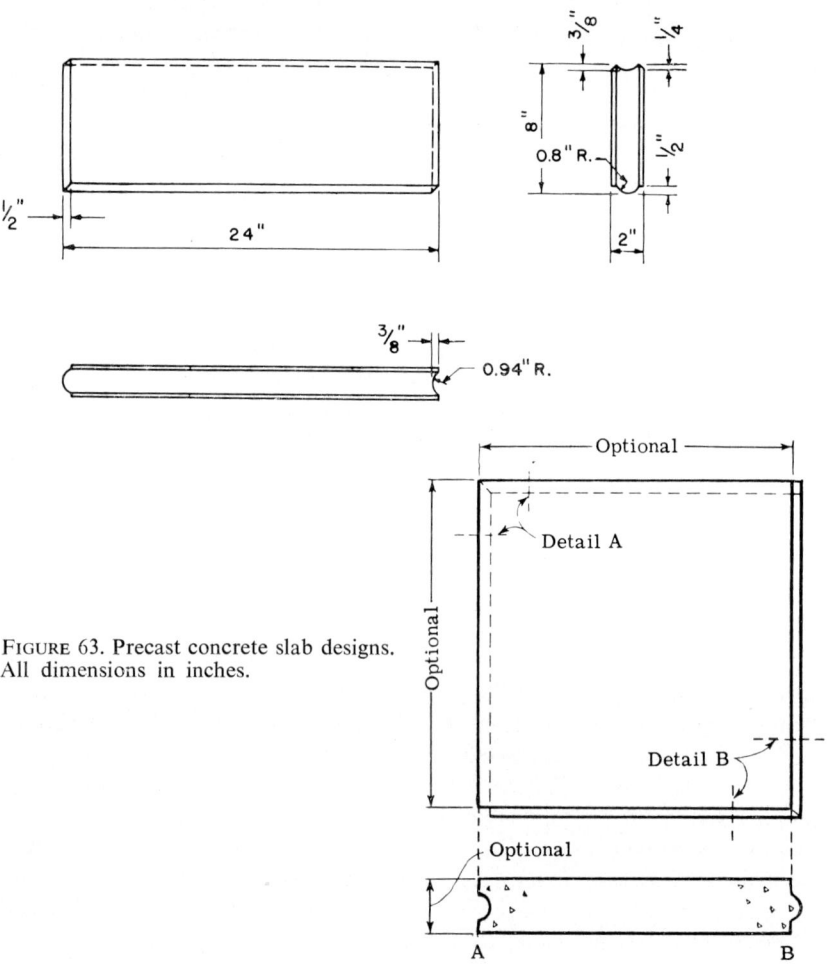

FIGURE 63. Precast concrete slab designs. All dimensions in inches.

FIGURE 64. Lining an irrigation canal with large concrete slabs on polyethylene sheeting, Hunger Steppe, Uzbekistan, U.S.S.R.

It is also to be noted that precast concrete slab linings in which all joints are sealed by a sealant such as bituminous mastic will be flexible enough to absorb minor movements of the subgrade without damage.

Precast concrete slab linings are subject to deterioration on sharp curves, which therefore should be lined by another method, such as *in situ* concrete or pneumatically applied mortar. Similarly, canal beds are often lined with *in situ* concrete, while prefabricated slabs are chosen to line the side slopes. Side slopes are more susceptible to subgrade movements and consequently to cracking of *in situ* concrete. It should also be noted that prefabricated concrete blocks can be laid with ease on steep side slopes (say 1:1) where *in situ* concrete is not always easily kept in place.

The slabs or blocks may be individually cast in forms at any suitable place, preferably near the pit providing the aggregate, or mass produced in a production plant. The slab units may be stored at any convenient location until such time as the labour is available to place them or until the canal operations permit primary lining works or repairs. Mechanized production requires a vibro-compressing machine. Small machines weighing about one metric ton are commercially available for easy transport and installation at any production site.

If machine placing is possible, slabs of considerable size and weight may be used. Figure 64 shows the placement of heavy concrete slabs on a sublining of plastic sheeting in the U.S.S.R. This combination of concrete and plastic creates a highly watertight and durable hard-surface lining.

Considerable handwork is required for precast concrete slab lining, which makes this type of lining competitive only where labour is abundant and cheap, or in cases indicated above, or for small jobs and repairs where maintenance forces may be utilized.

FIGURE 65. Stages in the lining of a branch canal with manually fabricated concrete slabs, India:
(*a*) Filling the forms with concrete.
(*b*) Moulding the concrete slabs.
(*c*) Curing the concrete slabs.
(*d*) Curing and stockpiling.
(*e*) Lining site.

a

b

Examples from India. Figures 65*a–e* show the prefabrication of concrete slabs for lining an old branch canal near New Delhi with a bed width of 6.40 m (21 ft), a design depth of 3.40 m (11 ft), 1:1 side slopes and a design capacity of 34 m³/sec (12.00 cusecs). The slabs are 60 cm (2 ft) square and 5 cm (2 in) thick. The concrete mix, prepared by machine mixer, consists of one part cement in three parts sand and six parts gravel (by volume). One mason and two labourers are able to produce around

c

d

e

200 slabs per day. One mason and four labourers can place about 130 m² (1 400 ft²) per working day.

Figure 66 shows a mould developed by Harmon (C90) for precasting small-size sections for farm irrigation channels which has been used successfully in Rajasthan, India. A concrete mixture of one part cement in two parts sand and three parts gravel is suitable. The gravel size should not exceed about 1 cm. The form is made of 2.5 cm (1 inch) thick hardwood. A 2.5 cm thick concrete layer is first placed to form the bottom and thoroughly tamped. The inner form is then placed over the concrete bed and the side gaps are filled with concrete, which is also thoroughly tamped. The mix must be relatively dry to allow early removal of the form for re-use. The cast concrete section has to be protected from the sun and wind for a day and then must be cured for at least one week. Consideration should be given to the use of metal rather than wooden forms, a slight inclination of the long sides toward the centre for easier removal of the form, a sliding guide for the inner form and other modifications.

The size of the section will depend on the volume of the irrigation stream to be conveyed; the size shown in the sketch is suitable for up to 30 l/sec (1 cusec).

After laying the sections, their joints are filled with cement mortar or any suitable asphalt-based sealing compound. The costs for this type of lining ranged from $0.40 to $0.70 per metre of channel in 1972.

SOIL-CEMENT LININGS

Soil-cement linings are constructed with mixtures of sandy soil, portland cement and water which harden to a concrete-like material. For the construction of soil-cement linings two general methods are in use: (1) the dry-mix method and (2) the plastic-mix method.

Dry-mix soil-cement

This type, which is called standard soil-cement or compacted soil-cement, is commonly mixed in place and compacted with the moisture content of the mix at or just above the optimum as determined by the Proctor test. Lining thicknesses commonly applied are 7.5 to 15 cm (3 to 6 in).

A typical sequence of dry-mix soil-cement lining construction of a farm reservoir, which applies as well to canal lining, is described by Rogers (C72): "... construction ... begins with spotting bags of cement at predetermined intervals on the bottom and sides of the reservoir. The bags are broken and the cement distributed by hand-raking or by me-

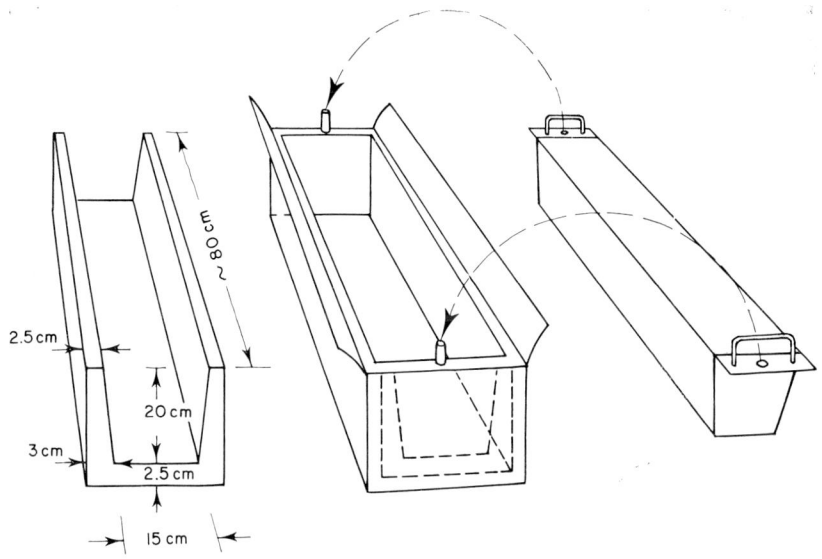

FIGURE 66. Form for casting small concrete channel sections.

chanical cement spreaders to a uniform depth. While the cement is being mixed into the soil with a rotary tiller, water is being added simultaneously to the mixture from a tank truck or hose. After the cement and soil are thoroughly mixed and the moisture content judged proper by the "hand-squeeze" method, the mixture is compacted by rubber-tired road compacters or heavily loaded trucks to approximately 10 cm (4 inches) thick. The soil-cement is usually cured by intermittent sprinkling for several days."

Mixing in place with travelling mixing machines has proved satisfactory on slopes not steeper than 4:1. For economic reasons side slopes of irrigation canals normally have to be much steeper. Therefore the mix should preferably be deposited directly on the side slopes from travelling mixers of the pugmill type or from equipment hauling it from stationary plants. An alternative method is the mixing of soil-cement for the side slopes on the previously finished and hardened canal bed and its transfer to the slopes by appropriate machinery or barrows. Entirely manual mixing and placing of dry-mix soil-cement has been carried out successfully in smaller canals.

The durability and watertightness of the dry-mix soil-cement lining is essentially dependent upon the soil used. Although other soils can be considered, laboratory tests indicate that for ease of mixing and placement and a low cement content the soil should be a well-graded, sandy, gravelly material with

100 percent passing a 3-inch (76-mm) screen;

55 percent or more passing a No. 4 (19-mm) screen;

10 to 35 percent passing a No. 200 (0.074-mm) screen.

If the soil is poorly graded or lacking in fines, the cement content should be higher, which also increases the cost. Since soils excavated from the canal are not necessarily the most suitable, soils located within a reasonable haul should be investigated; in some cases importing soil from a distant source may be more economical than using local soil. (The determination of the cement content is discussed under *Methods of testing soil-cement for canal linings* on page 112.)

Dry-mix soil-cement compaction should be completed within 1 hour after the soil-cement is spread. The moisture content of the material must be at or very near optimum when compaction begins. Compaction must be sufficient to obtain specified density. At the start of the project, in-place density tests are made of compacted soil-cement. From these tests the required number of passes of each type of compaction equipment can be ascertained. Thereafter, during construction, a minimum number of density tests are made daily. One agency in the U.S.A. tests every 500 yd^3 of soil-cement and makes at least two tests daily.

After the dry-mix soil-cement mixture has been compacted to the minimum specified density, curing should start immediately. Proper curing is just as important with soil-cement as it is with concrete. Improper or no curing can result in excessive shrinkage cracking, reduced durability, increased permeability, and lower strength. A continuous water cure is the most effective, followed by curing membranes such as proprietary pigmented compounds, bituminous emulsions, or cut-back oils. Even wet soil has been used successfully. The canal should be filled with water as soon as practicable. If the soil-cement can be kept at or near saturation, change in volume is minimized. Shrinkage cracks also will be largely eliminated or, if formed, will be tightly closed hairline cracks. Joints are not provided in dry-mix soil-cement linings.

Plastic soil-cement

Plastic soil-cement has higher water and cement contents than dry-mix soil-cement and a consistence comparable to that of concrete used for slipform lining. The soil is mixed with cement and water in a paver or mixer travelling along the canal or in a stationary plant. The mix is then poured by hand or by slipform on the subgrade to produce the lining. Lining thicknesses range from 7.5 to 15 cm (3 to 6 in). It is recommended that joints similar to those of concrete linings be provided. As an example, expansion joints may be placed every 30 m (100 ft) and contraction joints spaced on 7.5 m (25 ft) centres.

Tests with plastic soil-cement linings have been carried out in the Guararé Pilot Scheme of the La Villa River Project, Panama. Soil-cement ratios of 8 to 1, 11 to 1, 12 to 1 and 14 to 1 were tested. The lined canal had a base width of 1 m, a depth of 0.83 m, and 1.5:1 side slopes. Figure 67 shows the hand placing of the 7.5 cm (3 in) thick lining, using an 11:1 mix. Figure 68 shows the same canal under operation. The linings are said to perform well, but so far no comparative data are available.

FIGURE 67. Hand placing of a soil-cement lining approximately 7.5 cm thick in the Guararé Pilot Scheme, La Villa River Irrigation Project, Panama.

FIGURE 68. Completed reach of the canal shown in Figure 67.

Methods of testing soil-cement for canal linings

Test criteria and procedures cited here refer mainly to soil-cement for road paving, but are also commonly used for soil-cement canal linings.

Methods and procedures for testing soil-cement are thoroughly discussed in the *Soil-cement laboratory handbook* (C21), from which most of the following information has been extracted.

The composition of soils varies considerably, and this affects the degree to which they react when combined with portland cement and water. The way a given soil reacts with cement is determined by simple laboratory tests made on mixtures of cement with the soil. The amount of laboratory testing required for a given project depends on the requirements of the constructing agency, the number of soil types encountered, the size of the job and similar factors.

On major projects, for example, detailed tests are generally required, and the minimum cement content that can be used safely is determined for each significant soil type on the job. The cost of laboratory tests for major projects is quite small in comparison with the total costs.

On smaller projects, particularly where testing facilities and manpower are limited, it is sometimes considered advantageous to conduct only enough laboratory tests to determine a safe, but not necessarily minimum, cement factor that can be used for construction. For such cases the short-cut test method described below is adequate.

For very small projects where laboratory testing facilities are not available or detailed testing is not feasible or practical, a quick and very simple test procedure that involves moulding and inspection of specimens has been used successfully. It provides a safe cement factor but one that may be appreciably higher than the minimum for adequate hardness.

In all cases the tests are performed to determine three fundamental requirements for soil-cement:

1. How much portland cement is needed to harden the soil adequately?
2. How much water should be added?
3. To what density must the soil-cement be compacted?

Detailed test method. The detailed test method includes moisture-density, wet-dry, and freeze-thaw tests. (See the *Soil-cement laboratory handbook*.)

Short-cut test procedures for sandy soils. Short-cut test procedures have been evolved to determine adequate cement contents for sandy soils. Data and charts developed from previous tests of similar soils are utilized to eliminate some tests and greatly reduce the amount of work required.

The only laboratory tests required are a grain-size analysis, a moisture-density test and compressive-strength tests. Relatively small soil samples are needed, and all tests except the 7-day compressive-strength tests can be completed in one day.

While these procedures do not always give the minimum cement factor that can be used, they provide a safe cement factor generally close to that indicated by the detailed test method. The procedures are finding wide application by engineers and builders and may largely replace the standard tests as experience in their use is gained and the relationships are checked. The charts and procedures may be modified if necessary to conform to local conditions.

Rapid test procedure. A rapid method of testing soil-cement has been used successfully for emergency construction and for very small projects where more complete testing is not feasible or practical. It involves moulding and visually inspecting several specimens that cover a wide range of cement contents — for example, 10, 14 and 18 percent. After at least a day or two of hardening, the specimens are inspected by "picking" with a relatively sharp-pointed instrument and by sharp "clicking" of each specimen against a hardened object such as concrete to determine the relative hardness. If a specimen cannot be penetrated more than 3 to 6 mm ($1/8$ to $1/4$ in) by picking and if it produces a clear or solid tone upon clicking, an adequate cement factor is indicated.

Even an inexperienced person can soon differentiate between satisfactorily and unsatisfactorily hardened specimens and will be able to select a safe cement content to harden the soil.

Estimating cement requirements. The following information will aid the engineer in estimating cement requirements of the soils proposed for use and in determining what cement factors to investigate in the laboratory tests.

As a general rule, it will be found that the cement requirement of soils increases as the silt and clay content increases, gravelly and sandy soils requiring less cement for adequate hardness than silt and clay soils.

The one exception to this rule is that poorly graded, one-size sandy materials devoid of silt and clay require more cement than do sandy soils containing some silt and clay.

In general, a well-graded mixture of stone fragments or gravel, coarse sand, and fine sand either with or without small amounts of feebly plastic silt and clay material will require 5 percent or less cement by weight. Poorly graded one-size sand materials with a very small amount of nonplastic silt, typical of beach sand or desert blow sand, will require about 9 percent cement by weight. The remaining sandy soils will generally require about 7 percent. The non-plastic or moderately plastic silty soils generally require about 10 percent cement by weight, and plastic clay soils require about 13 percent or more.

Table 13 gives average cement requirements for a number of different materials and special types of soil which have been used successfully in soil-cement construction to replace soils that required high cement contents for adequate hardening.

TABLE 13. — AVERAGE CEMENT REQUIREMENTS OF VARIOUS MATERIALS

Material	Percent by volume	Percent by weight
Caliche	8	7
Chat	8	7
Chert	9	8
Cinders	8	8
Limestone screenings	7	5
Marl	11	11
Red dog	9	8
Scoria containing plus No. 4 material	12	11
Scoria (minus No. 4 material only)	8	7
Shale or disintegrated shale	11	10
Shell soils	8	7
Slag (air-cooled)	9	7
Slag (water-cooled)	10	12

Cost and performance

Little information is available on the cost of soil-cement linings. Johnson (C53) states: " Soil-cement linings cost only about half as much as concrete linings. It is the type of job that lends itself to station operations using their own forces."

Some data on executed soil-cement linings are compiled in Table 14.

TABLE 14. — EXAMPLES OF SOIL-CEMENT LININGS

Project	Specification of lining material (percentage per weight)	Findings
Main Canal, W.C. Austin Project, Oklahoma, U.S.A. Experimental installations (A1)	*Dry-mix* Soil: 100% passing No. 4 sieve (19 mm) 60% passing No. 200 sieve (0.074 mm) Portland cement: 10-12% Thickness: 10-15 cm (4-6 in) Installed in 1945	After 12 years of service, in poor condition.
As above	*Plastic* Soil: as above Portland cement: 16% Thickness: as above Installed in 1945	After 12 years of service, in fair condition.
As above	*Plastic* Soil: poorly graded silty fine sands Portland cement: 4 different sections containing 11.1 to 22.2% Thickness: all sections 8 cm (3 in) Installed in 1947	After 10 years of service the lining with 11.1% cement content was badly deteriorated. The sections with 14.5 to 22.2% cement content were in fair to good condition and still effective in reducing seepage losses.
Canal on the Boise Project, Idaho, U.S.A. Experimental reach (A1)	*Plastic* Portland cement: 14.3% Thickness: not quoted Installed in 1948	After 14 years of service, still in very good condition.
Storage reservoir, Port Isabel, Texas (C72)	*Dry-mix* Portland cement: 12% Soil moisture content: 18% Thickness: 10 cm (4 in) Constructed in 1945	After 23 years of service, the soil-cement was still performing satisfactorily and the maintenance had been negligible
Drainage canals, Calverton, Long Island, New York (C53)	*Dry-mix* (mixed in place) Soil: sand of open texture with no binder Portland cement: 14.5% Cured with asphalt emulsion Thickness: not quoted	In good condition after 4 years, except for some deterioration on steep slopes. (Cause: non-uniform mixing of cement and soil and inadequate compaction.)

TABLE 14. — EXAMPLES OF SOIL-CEMENT LININGS (*concluded*)

Project	Specification of lining material (percentage per weight)	Findings
Drainage canals, U.S. Naval Auxiliary Air Station, Whiting Field, Florida (C53)	*Plastic* Soil: sandy with 10-20% silt, scarcely any clay Portland cement: 13% Thickness: 13 cm (5 in) Cured with wet earth or curing compound	Very effective after 8 years of service. Very little, if any, noticeable deterioration has occurred.
Small irrigation canals of the Eden Project, Wyoming, U.S.A. Field tests for the evaluation of dry-mix soil-cement linings (C62)	*Dry-mix* (machine mixed) Soil: sandy with high sulphate content Portland cement: 9% (design); 5.3-5.8% (actual) 12% (design); 6.5% (actual) Thickness: not quoted	Lauritzen states: "Seepage losses associated with intermittent use seldom justify lining field ditches on the basis of the value of the water saved in this or other areas. Ditch stability may be all that should be sought. If this is assumed, soil-cement — even that of poor quality — has been reasonably effective for a period of over 10 years."

ASPHALTIC CONCRETE LININGS

Asphalt is a petroleum product which, when mixed with sand and gravel, is used as a liner in much the same way as concrete made from portland cement. Asphaltic concrete [1] linings when properly constructed are comparable to portland cement concrete linings in many respects. The expected service life of asphaltic concrete linings, however, is shorter than for linings constructed of portland cement and ranges between 10 and 20 years. The maximum permissible velocity should not exceed 1.50 m (5 ft) per second. It will be necessary to include subgrade sterilization as an integral part of the construction operation to prevent plants from growing and penetrating the lining, although this may not apply to canals constructed in desert areas. The advantages as compared with portland cement concrete linings include the possibility of placement

---

[1] In this publication the term "asphaltic concrete" is employed to describe any mixture of solid or semisolid bitumenous products obtained from the distillation of petroleum and any kind of aggregate.

during freezing weather, its better adjustment to subgrade changes, and the possibility of using slightly poorer quality aggregate.

Any overall advantage in the use of asphaltic concrete for canal lining would be contingent upon a considerable price differential between asphalt and portland cement in favour of asphalt and the adaptability of local aggregates to asphalt construction.

Hot-mixed asphaltic concrete

This is a carefully controlled mixture of asphalt, mineral filler and graded aggregate which is mixed, placed and compacted at elevated temperatures. Mix designs for canal lining purposes are very similar to dense-graded mixes for highway surfaces. They differ in having a higher mineral filler and asphalt content. Two recommended gradings for canal lining purposes are given in Table 15.

TABLE 15. — SUGGESTED MIX COMPOSITIONS FOR DENSE-GRADED ASPHALTIC CONCRETE LININGS

Sieve size U.S. Standard	For minimum thickness of 3.8 cm (1½ in)	For minimum thickness of 2.5 cm (1 in)
	.... Percent passing	
³/₄ inch	100	
½ inch	95 - 100	
³/₈ inch	—	100
Number 4	60 - 80	90 - 97
Number 8	45 - 60	70 - 85
Number 30	28 - 39	42 - 52
Number 100	16 - 25	20 - 28
Number 200	8 - 15	10 - 16
Asphalt cement percent by weight of total mix	6.5 - 8.5	7.5 - 9.5

A binder having a penetration of 60 to 70 is preferred for use in canal lining. This relatively hard penetration grade produces linings that are more resistant to the destructive action of water, extremes of weather and penetration by vegetation. They are stable on side slopes, yet they retain sufficient flexibility to conform to slight deformations of the subgrade. The recommended lining thickness can be determined from Figure 26 (page 66).

The watertightness of the lining is highly dependent on the degree of compaction. This should be specified to equal at least 95 percent of the laboratory density attained by the Marshall method, or a comparable degree of compaction. Such compaction produces a lining with less than 5 percent air voids, making it waterproof and weatherproof.

Asphaltic concrete lining is especially well adapted to smaller canals where placement can be accomplished by subgrade-guided slipforms. The asphaltic concrete is mixed on or near the job, and hauled by trucks to the point where it is to be laid. Slipforms used for asphaltic concrete lining are similar to those for producing cement concrete linings. The specified degree of compaction is attained by smooth-wheel rollers, vibratory rollers or weighted and heated ironing screeds.

The initial placement cost of hot-mixed asphaltic concrete lining varies considerably, but is generally not lower than that for portland cement concrete linings.

Little hand labour is used in the processing of hot-mix asphaltic concrete lining. Therefore, the lining job must be of considerable magnitude to absorb the cost of aggregate production, moving and setting up the hot-plant, and the use of trucks, slipform and compaction equipment.

In areas with surplus hand labour and a shortage of foreign exchange, hot-mix asphaltic concrete linings are probably not economical.

Figures 69a,b show a machine used in lining 100 km of canal in the Federal Republic of Germany (Wasser- und Schiffahrtsdirektion, Hamburg) with asphalt concrete in 1973. The canal has a bed width of 26 m, a design water depth of 4.65 m and 1:3 side slopes. While the lining thickness of 16 cm asphalt concrete on a layer of 6 cm asphalt-sand mix (Figure 70) is not applicable to irrigation canals, the lining method and concrete mix as such may be considered, selecting the thickness according to Figure 26 (page 66).

The mix composition is as follows:

	Percent
— Asphalt (bitumen B80, according to German standards)	6.3
— Mineral filler	9.5
— Sand	51.2
— Gravel	33.0
	100.0

The mix was conditioned by a maximum permissible air void of 3 percent.

FIGURE 69. Two views of an asphaltic concrete slope-lining machine on the North-South Navigation Canal, Fed. Rep. of Germany.

FIGURE 70. Test core hole in asphaltic concrete lining.

Average placement capacity of the machine was 2 000 metric tons of mix in a 10-hour working day.

Cold-mixed asphaltic concrete

This material is similar to the hot-mixed type in that well-graded aggregates and asphalt are mixed and compacted in place. However, it is necessary to cure cold mixes, which requires time and favourable weather conditions. Since cold mixes tend to exhibit low erosion resistance and stability for an appreciable period of time, while construction costs are not significantly lower than for cement concrete lining, their use for irrigation canal lining is not recommended.

PREFABRICATED ASPHALTIC LININGS

Materials similar to asphaltic concrete have been prefabricated to avoid the use of hot materials on the job, which requires skilled personnel and special equipment.

Thin prefabricated asphaltic materials are employed as buried membranes. Thick prefabricated asphalt slabs have been developed for use as exposed linings. Owing to its relatively short service life and high transportation costs, this type of lining may not be economical for most uses.

BRICK LININGS

Brick linings have many of the advantages and disadvantages inherent in linings constructed of precast concrete slabs. Brick linings, however, can be used to advantage in areas where

— an abundance of inexpensive manual labour is available;
— materials for brick are available near the canal sites;
— materials and machinery for other suitable types of lining are not available at a competitive price.

Examples of brick linings

Economic preconditions for extensive use of brick linings are found in the plains of India and Pakistan. In the past, the most frequently used conception was the reinforced double-tile lining. This type is illustrated by Figure 71 showing the lining used in a reach of 80.5 km (50 miles) of the Haveli Canal in the Punjab, completed in 1938. The canal, which has a design capacity of 170 m³/sec (6 000 cusecs), has a bed width of 25.6 m (84 ft), a depth of 3.6 m (12 ft), and side slopes of 1:1. Seepage measurements in two reaches of the canal with the ponding method gave the following results:

Reach 1: 26.4 l/m²/24 hours (0.1 cusec per million ft²)

Reach 2: 66.0 l/m²/24 hours (0.25 cusec per million ft²)

In later projects the brick lining was placed on a carefully compacted subgrade instead of being reinforced. (Detailed descriptions of construction of double-tile lining are found in A15 and A16.)

FIGURE 71. Double-tile lining, Haveli Canal, India.

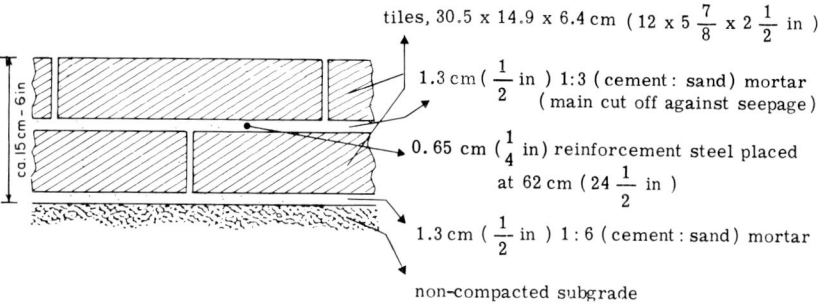

In recent years the so-called single-tile lining has proved more economical, provided that the subgrade is well compacted and that arrangements are made to release hydraulic uplift pressure when this may occur. Figure 72 shows two designs for a single-tile lining, both of which have been constructed and tested on the Sirhind Feeder in India.

This canal has a discharge capacity of more than 133 m^3/sec (4 700 cusecs), is 24.4 m (80 ft) wide, 5.2 m (17 ft) deep, and has side slopes of 1.25:1. The test results confirmed the efficiency of the single-tile lining regarding permeability, structural strength, ease of construction and satisfactory operation (C85).

Construction aspects

Clay brick is generally a very porous material. The brick layers, therefore, hardly play any part in preventing seepage losses. The watertightness in brick linings is solely due to the plaster sandwich. The bricks only form a skeleton to hold the plaster or function as a protective cover.

The plaster normally has a thickness of from 2.0 to 2.5 cm (3/4 in to 1 in), depending on the head of water in the canal, and is made of one part cement on three parts sand. The sand used must not have a fineness modulus of less than 1.2. The percentage volume of silt should not be allowed to exceed 6 percent (A3).

It has been common practice in India and Pakistan to replace 25 to 30 percent of the portland cement either by Surkhi, made by grinding

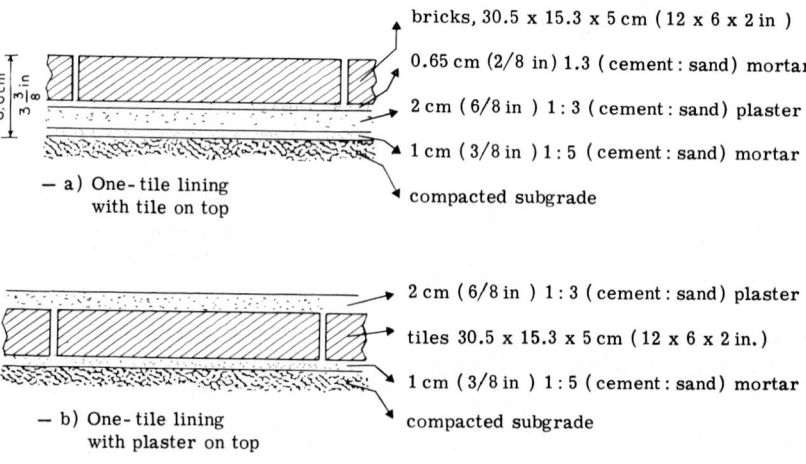

FIGURE 72. Two types of one-tile lining, Sirhind Feeder, India.

burnt bricks, or by a locally exploitable lime called Kandar. This not only has economical advantages when portland cement is in short supply, but Surkhi has also proved to increase the watertightness considerably because of reduced temperature cracking (A3, A6).

Bricks are manufactured from earth in which the salt content does not exceed 2 percent. The clay content should range between 10 and 20 percent. The bricks are thoroughly soaked before use (A3). Many shapes and sizes of bricks are in use but little information is available to show the superior value of one type over others.

No earth pressure or hydraulic uplift pressure should be exerted on brick linings. The side slope, therefore, should always be less than the angle of repose of the subgrade. In some projects, slopes of 1:1 have failed and have been flattened to a slope of 1.5:1. Uppal (C85) recommends that when the permeability of the subgrade soil is between 1 and 15 m (3 and 50 ft) per year, drainage arrangements may be required. Figure 73 shows a simple type of drainage outlet from a sand filter behind a brick lining.

When the water level in the canal drops because of sudden upstream closures or other reasons, the outlet will drain the excess water present in the adjacent soil and relieve the pressure behind the lining.

Asphaltic or synthetic membranes in brick linings

When bitumen is available at competitive prices, membranes of this material or bitumen-sand can be placed between or under a brick lining.

FIGURE 73. Drainage arrangment for brick lining in soils with low permeability.

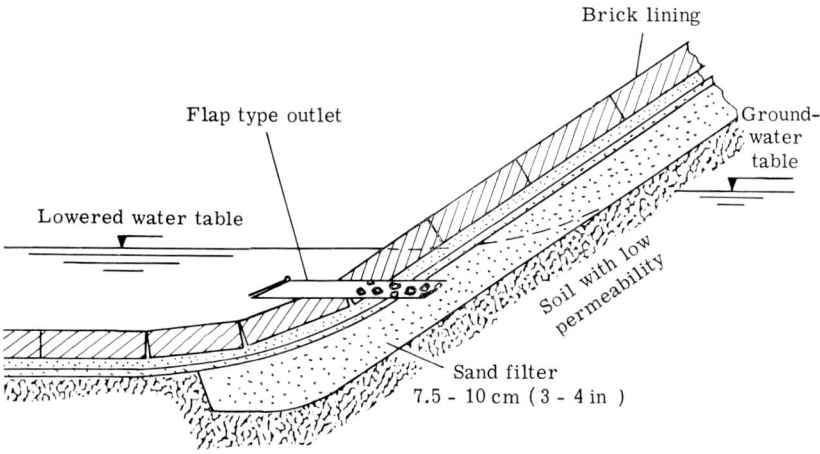

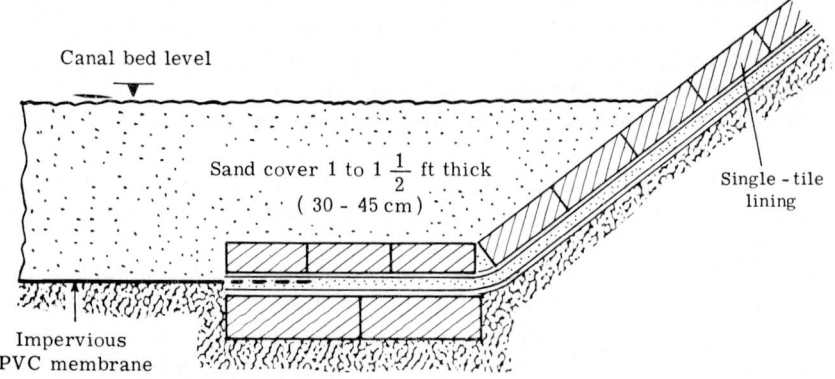

FIGURE 74. Combined tile and buried membrane lining.

This may replace the concrete plaster normally employed or may be used in addition to the plaster to increase the watertightness of the lining.

Laboratory tests and field trials in Pakistan (C67) have shown that a 3.2 mm ($^1/_8$ in) coat of bitumen above a conventional 2 to 5 cm thick 1:3 cement/sand plaster and overlaid by a protective brick layer makes the lining practically impervious. In order to provide adhesion between the coat and the bricks, dry sand is spread over the hot bitumen.

Uppal, Midha and Singh (C85) report on the development of another combination type of lining in India. This lining consists of a low-cost impervious membrane of plastic or synthetic rubber buried under a sand cover on the bed and a brick lining on the sides. The method is best suited to lining existing canals with large bed widths (Figure 74).

A reach of the existing unlined Rasulpur distributary (Q = ca. 5 m³/sec = 174 cusecs) was lined in 1963/64 with a combination lining. The cost of lining 1 km of canal amounted to about U.S.$7 000 (1967 exchange rate). It was originally proposed that this distributary be replaced by a new canal with a conventional brick tile lining. The cost estimate for this alternative was $14 500 per km.

A drawback of this type of lining is that the velocity must be decreased to prevent erosion of the sand cover.

Clay tiles in small channels

The use of semicircular clay tiles for lining farm irrigation channels is reported by Michael, Reeser and Knierim from India (C90). The tiles are moulded by hand using a wooden form made in halves. The size is limited to 30 cm (12 in) in diameter and 37 cm (15 in) in length; larger tiles are distorted or collapse owing to their weight when turned out of the form. The tiles are provided with a bell joint (collar) on one end. A village potter of ordinary skill can manufacture the tiles. The tiles are baked in a kiln in the same way as earthen water pots or roof tiles. Larger farm or interfarm ditches can be lined by using semicircular tiles at the bottom and rectangular tile slabs as extensions. A common size of rectangular tile slab for hand moulding is 37 × 30 cm (15 × 12 in). The tiles are joined with cement mortar. To prevent damage, men and animals should not be permitted to walk in the channel. When protected from external damage, this clay-tile lining will last for several years.

STONE LININGS

Linings of stone masonry have been employed in areas where suitable materials, such as sandstone or basalt, are available. The construction of stone linings is relatively slow, and labour is the major expense, as the stones are placed by hand.

Seepage losses may be very high if the stones are not mortared. Non-mortared stone linings have proved advantageous where erosion control is the main reason for lining, as on steep slopes. Figure 75 shows a stone-lined canal in Mauritius with a capacity of about 1.1 m^3/sec. Initially, in 1952, the lining was placed without mortar joints, but excessive seepage losses made it necessary to seal the joints with mortar. Some reaches had to be relined with reinforced concrete. Figures 76 and 77 show other examples from Hawaii and the Yemen Arab Republic.

On a project in Peru (A3), it was computed from actual lining expenses that a stone masonry lining was about 40 percent cheaper than a concrete lining for a canal with a capacity of 50 m^3/sec. Yet labour costs for the stone lining were about double those for the concrete lining.

Figure 78a shows a canal in India being lined with "Katla" stone. This material is a sedimentary rock — mostly sandstone — which is abundant in the Chambal area, Rajasthan. It is usually quarried in large slabs up to 30 cm thick and is then split with wedges down to the thickness required for use, the minimum being about 2 cm.

The main problem in using this material for canal lining is that the coefficient of expansion at high temperatures causes cracking at the joints if these are not properly made. Due to their low unit weight the slabs are also susceptible to damage by high water velocities and turbulence, as shown in Figure 78b. In the area of Chambal in Rajasthan the stone slabs have been used on clay soils and it has been the practice not to place sand layers under the slabs. Expansion and contraction of

FIGURE 75. Stone-lined irrigation canal with mortar joints, Mauritius.

FIGURE 76. Masonry lining of lava rock mortared together, Hawaii.

FIGURE 77. Stone-lined canal for conveying water from a well, near Dhamar, Yemen Arab Republic.

FIGURE 78a. Placing "Katla"-stone slabs in a main canal, Chambal Project, India.

FIGURE 78b. "Katla"-stone lining damaged by high velocities and turbulence at a pipe outlet.

the clay subgrade have consequently caused frequent failure of these linings; thus sand cushioning and properly constructed joints have been suggested for the construction of reasonably stable and impervious linings.

It would seem advantageous to provide a 0.2 mm thick polythene sheet (see page 129) underneath the stone lining, which would have a protective function only, while the problem of leakage through the joints would be taken care of by the impermeable membrane.

Exposed membranes

Exposed membranes include thin membranes of asphalt, plastics and synthetic rubber. These materials are also used much more effectively for buried linings. Their low permeability, when combined with the strength of soil or other base material, will prevent seepage of water. While the membranes differ somewhat from one another, they differ completely from rigid-type linings in that they are not expected to provide only impermeability, not structural strength.

Tests in the United States (C69) indicate that plastic materials placed as exposed membranes deteriorated from sun, weather and erosion after two to four years. Experimental installations in various parts of the world confirm these findings. Damage is also commonly caused by weed puncture and weed burning, livestock traffic, maintenance equipment, rodents and vandalism. Even theft has been a problem in some areas.

Butyl-coated fabrics, particularly nylon, perform much better than plastics but are more expensive.

Because of short life the economic use of exposed membrane linings is limited to special cases, as for temporary emergency linings and short sections, or where there is little hazard of mechanical damage or prices are competitive owing to special transport and other conditions.

INSTALLATION

The installation of the membrane is quite simple after the subgrade has been prepared. All rocks, stones, roots and other sharp objects that might puncture the membrane should be removed or covered with a few inches of sand or soil. A perimeter trench, about 30 cm deep, should be prepared above the water and wave line for anchoring the upper edge of the membrane. The slope can be as steep as the natural subgrade soil permits. A Manning's coefficient of 0.012 may be applied for hydraulic calculations of canals lined with exposed membranes.

For the preparation of the subgrade and placing of a black polyethylene film on some existing unlined ditches, Corry and Scott (C38) give the following figures on manpower requirements. A rate of 44 m (143 ft) per man per hour was achieved under the conditions of a straight canal, uniform in cross section and free from mature weeds and trash. The rate dropped to 6.7 m (22 ft) per man per hour if soil which sloughed into the canal had to be removed and the film had to be fitted to a very irregular cross section.

Covered membrane linings

Experiments and practice show that membranes such as sprayed-in-place asphalts, prefabricated asphaltic materials, plastics, synthetic rubber, clay and bentonite effectively control seepage over a considerable period of time when covered. The need for a protective cover became apparent after early trial installations showed that the exposed membranes had little resistance to the various field hazards.

The use of buried membrane canal lining for rehabilitation on operating projects has a distinct advantage over most other types of linings, since this type of construction may be employed during cold or wet weather, such as is frequently encountered during the non-irrigation season.

GENERAL DESIGN CONSIDERATIONS

Subgrade

The canal section must be excavated to provide for the required water prism, plus the cover material. Generally the native soil subgrade should be disturbed as little as possible in the excavation or reshaping operation. The subgrade should be smooth and firm. Dragging the subgrade with a heavy marine-type chain or tractor track has proved to be a rapid and effective method for smoothing canal subgrade. Sterilization may be necessary to prevent vegetation from penetrating the lining. Compaction of fill section is not as important as for rigid lining. The steepness of slope is dependent upon the cover material used and is normally much flatter than the angle of repose of the subgrade material. The U.S. Bureau of Reclamation recommends a maximum slope of 2:1 for asphalt or plastic membranes. If, however, the cover is composed of relatively unstable material, such as uniformly graded sands, fine gravels or silty sands, a flatter slope must be used. Lauritzen (C59) and Staff (C75) recommend a maximum slope for earth-covered membranes of 3:1. This is confirmed by the American Society of Agricultural Engineers (ASAE) Recommendation R340 of 1970.

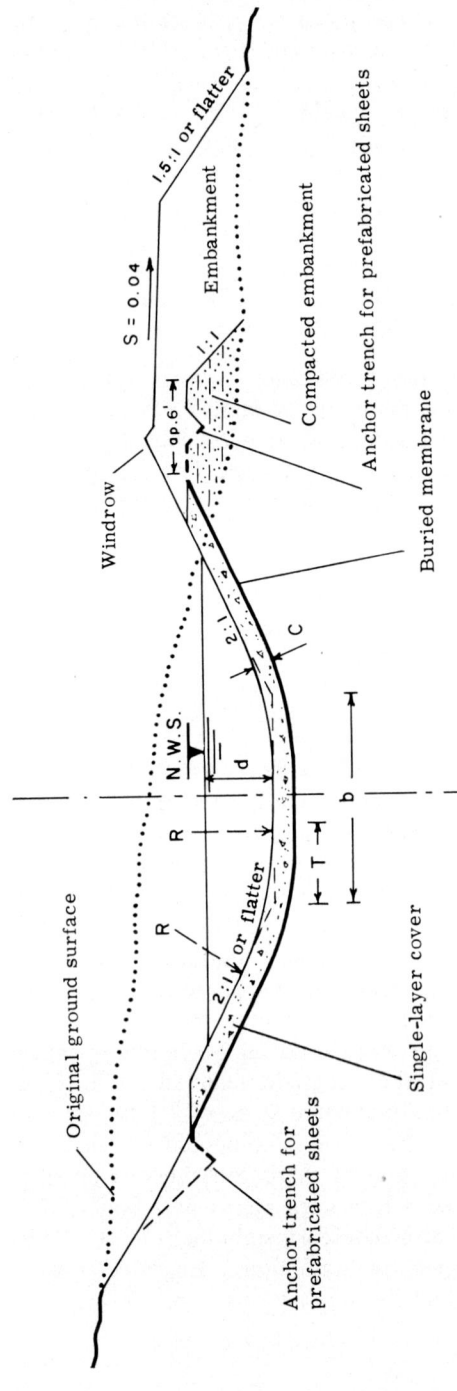

FIGURE 79. Typical section of buried membrane lining installation.

$C = 10$ in $+ \dfrac{d}{12} \times$ where C and d are expressed in inches.

Ratio b:d in feet (usually 4:1).

b	2	5	10	20	30	40	50
T	1	1.5	3.5	6.5	8.7	10.5	12.0

Slope of 2:1 is a borderline case; 3:1 is normally recommended.

A typical cross section of a buried membrane lining is illustrated in Figure 79.

Protective cover

Since earth excavated from the canal section is usually the least costly cover material, it is the most frequently used. The thickness of the layer depends on the erosion resistance of the material and local conditions, such as type of cleaning equipment, amount and type of animal traffic within the canal section, and localized scour, particularly at curves and structures. For a clayey gravel or an equally erosion-resistant material the U.S. Bureau of Reclamation recommends a minimum thickness of one twelfth the normal water depth plus 25.4 cm (10 in). Should the local material be fine-grained and noncohesive, a greater total thickness may be required.

Tests have been carried out in India (C84) to determine the thickness of soil cover for protecting a membrane against damage from animal traffic. From these tests it was concluded that maintaining 30 cm of "sound cover" will protect the membrane with a fair margin of safety against livestock damage.

The ASAE (C32) recommends a minimum thickness of soil cover for flexible membranes of 15.2 cm (6 in). The bottom 7.6 cm (3 in) next to the membrane should not be coarser than silty sand. Where a flexible membrane is to be buried for protection against livestock or mechanical damage, a minimum of 23 cm (9 in) of cover is required.

Some difficulty has been experienced with the cover material slipping down the slope when two layers (gravel over earth) have been used. If the erosion resistance of the local material is insufficient, it is preferable to mix gravel into the earth and place the mixture in one layer. The slippage of the cover material on the membrane is a problem particularly during drawdown. If the cover is not free-draining without loss of fines, rapid fluctuations in the water surface must be avoided. The magnitude of permissible drawdown may be illustrated by the results of a test conducted by the U.S. Bureau of Reclamation (C18). The cover material used was a mixture of sand and gravel (95 percent larger than 0.074 mm and 25 percent larger than 2 mm). It was placed 40 cm thick on a sprayed-in-place asphalt membrane with a slope of 2 to 1. It was found that this cover could withstand drawdown rates up to 60 cm per hour without sloughing.

When subject to surface waves, the cover material used in the above-mentioned test eroded rapidly at first, but soon became rather firm and armoured with the coarse material. In larger canals a beach belt of gravel is sometimes installed to increase safety against wave action.

In designing a buried membrane lining, it should be borne in mind that the cover material limits the permissible velocity. This will usually be somewhat less than that in an unlined canal of similar soil (A1). The ASAE (C32) recommends that design stream velocity for buried flexible membrane canal linings should not exceed 0.9 m/sec (3 ft/sec). Maximum safe non-erosive velocities for different soils are shown in Tables 3 and 4. For freeboard with buried membrane linings, see Figure 16 on page 53.

The placement of the protective cover should begin as soon as the membrane lining is judged to be properly positioned, jointed and repaired, if necessary. The cover material can be placed with care by draglines, conveyors, trucks, or other means, preferably starting at the toe and working upslope. It is not necessary to compact the cover material by rolling; however, dragging is usually required to attain the required finished shape and uniform thickness.

BURIED SPRAYED-IN-PLACE ASPHALTIC MEMBRANES

Buried asphaltic sprayed membranes represent by far the greatest volume of membrane lining in service. To date there has been no report of the aging of asphalt having been considered a contributor to failure of the membrane. Failures have been attributed to too thin a membrane, to penetration by weeds and to rupture due to movement of the cover material.

The asphalt that gives the best performance is a special high-softening-point asphalt made by blowing with a phosphoric compound. In brief, the material has the following properties:

Flash point (Cleveland open cup), not less than 220°C (425°F).

Softening point (ring and ball), between 80°C (175°F) and 95°C (200°F).

Penetration at 25°C (77°F) (100 g, 5 sec), 50-60 scale.

Penetration at 0°C (32°F) (200 g, 60 sec), not less than 30.

Penetration at 46°C (115°F) (50 g, 5 sec), not more than 120.

Ductility at 25°C (77°F) (5 cm per min), not less than 3.5 cm.

Loss at 163°C (325°F) in 5 hours, not more than 1 percent.

Penetration of residue at 25°C (77°F) (100 g, 5 sec), not less than 60 percent of penetration before heating.

Bitumen soluble in carbon tetrachloride, not less than 97 percent.

FIGURE 80. Second application of asphaltic membrane with handborne spray bar, Columbia Basin Project, U.S.A. (USBR).

Asphaltic membranes are usually placed in thicknesses from approximately 5 to 8 mm (3/16 in to 5/16 in). This requires at least 5.5 litres of asphalt per square metre to assure a continuous waterproof layer. The average quantity applied is ca. 7 l/m^2. Rough surfaces require heavier applications. Sprinkling of dry, dusty subgrades prevents holidays and pinholes.

The application temperature of the asphalt should be from 175 to 210°C (about 350 to 410°F). Handsprays to apply the material are used occasionally, but long multiple spray bars which extend over the entire canal side or bottom (Figure 80) are favoured for more uniform application. A pressure of 3.5 kg/cm^2 (50 lb/in^2) is necessary. The high temperature of application will require the use of an auxiliary heater. Usually one application is sufficient for bottom placement, but two to three passes are required on the sides. In the United States, three passes with the distributor moving at 6.5 km/h (about 4 mi/h), using at least a 2.5-m bar, produced excellent results. The asphalt cools quickly and is soon ready for the application of cover material.

The handling of high-softening-point asphalt requires skill and organization of work to prevent its "freezing" in the hose lines or spray bars. By carefully coordinating the work, numerous contractors have been able to apply satisfactorily over 40 000 litres daily with a single crew (A3).

Cold-mix asphalt is not recommended for membrane lining.

Prefabricated Asphaltic Membrane Linings

This category of membrane comprises all thin asphalt-coated felts or fibre-mats, similar to those used in the roofing industry. Reinforcing materials used are jute, hemp, glass fibre and asbestos. The membranes have a thickness of appr. 3 to 6 mm ($1/8$ to $1/4$ in) and are made up in rolls of standard size.

Construction procedures similar to those used in the placement of the hot-applied membranes (see above) should be followed in subgrade preparation and placement of the protective cover. All joints of the membrane are usually lapped at least 5 cm. Adhesives used are hot asphalt cements, cold mastics or special cutbacks (see Figure 81).

Prefabricated asphaltic membranes have been developed for use on smaller canals or relatively short reaches of large canals; when skilled personnel and equipment for sprayed-in-place asphalt cannot be provided; where space limitations and other site conditions prevent the use of equipment for the sprayed-type membrane; in isolated areas, remote from a supply of the special asphalts required for the sprayed-type membranes.

Plastic and Synthetic Rubber Membrane Linings

Extensive use is being made of a variety of plastic and some rubber prefabricated membranes for canal lining. The most commonly used materials are plasticized polyvinyl chloride (PVC), polyethylene (PE) and butyl rubber. In recent years the puncture resistance of PE has been considerably improved. Yet PE remains the most susceptible to sun

FIGURE 81. Placing lightweight, buried, glass-fibre-reinforced, prefabricated asphaltic canal lining, Altus Project, Oklahoma, U.S.A. (USBR).

damage of the three materials mentioned. Considering both performance and cost, in the majority of practical applications PE may be the most economical material for buried flexible membrane canal lining.

The condition of materials that have been in place for 10 to 15 years or more, plus accelerated testing, indicate a long life expectancy for properly prepared materials. A 0.2 mm (8 mil) buried PE lining was installed in two reaches of an irrigation canal in Canada in 1955 and 1956 (B30). Seepage tests carried out immediately after lining and in 1965 showed the following losses ($m^3/m^2/24$ h):

	SEEPAGE LOSSES ($m^3/m^2/24$ h)	
	1955/56	1965
Reach 1	0.0046	0.0040
Reach 2	0.0049	0.0034

These results indicate that the linings were still effective after 9 and 10 years of service.

Materials

Table 16 can be used as a guide in selecting the proper membrane thickness relative to subgrade conditions. Membrane thickness should not be increased to compensate for improper subgrade preparation.

Installation

The linings are supplied in large pieces to minimize the amount of field joining, although joining does not require highly trained labour. Polyethylene and butyl are supplied in folded rolls, while PVC is generally accordion-folded in both directions so that the liner can be opened lengthwise from the back of a truck or from a fork lift. Both rubber and plastic sheeting should be installed in a relaxed state or with slight slack allowed in both directions. PVC and PE films should have about 5 percent slack in both directions.

All membranes need to be joined in the field, rather than only overlapped, to provide watertightness. PVC liners are quickly sealed to give rapid setting by injecting a small amount of adhesive between the two pieces, which are lapped 5–10 cm (2–4 in), and by gently smoothing the adhesive-covered area to effect a bond. A few minutes after the

TABLE 16. — CHARACTERISTICS OF COMMONLY USED POLYMER MEMBRANES

	Polyvinyl chloride (PVC)	Butyl (unreinforced)	Polyethylene (PE)
Tensile strength (minimum average)[1]	140 kg/cm^2 (2 000 psi)	84 kg/cm^2 (1 200 psi)	126 kg/cm^2 (1 800 psi)
Ultimate elongation (minimum average)[2]	300 percent	300 percent	500 percent
Available gauges[2]	0.20-0.85 mm (8-35 mil)	0.8-3.0 mm (32-125 mil)	0.15-0.5 mm (6-20 mil)
Available widths[2]	1.20-19.0 m (4-61 ft)	8.5-14.0 m (28-46 ft)	5.0-12.0 m (16-40 ft)
Method of joining[2]	heat, solvent or adhesive	adhesive	heat, tape or adhesive
Density[2]	1.25	1.25	0.93
Resistance to sun exposure	poor	good	very poor
Minimum thickness acceptable for general use:[1]			
Coarse soils	0.20 mm (8 mil)	0.76 mm (30 mil)	0.20 mm (8 mil)
Gravels	0.30 mm (12 mil)	0.76 mm (30 mil)	0.30 mm (12 mil)
Cost[2]	medium	highest	lowest

[1] According to ASAE Recommendations (C32). — [2] According to Staff (C75).

bond has been made, the large sections can be opened. Special adhesives that develop strength more slowly are used with PE and butyl. Tape can be used to join PE.

Installation is normally begun at the downstream end of the canal (Figure 82). Connections to pipes and other structures are quite easily made with adhesive and pieces of the sheeting. Seals to round structures (e.g., pipes) are made by opening a hole in the membrane somewhat smaller than the pipe, applying adhesive to the pipe, and forcing the small hole over the end of the pipe. This will turn up a collar on the pipe, thus making a watertight seal. The collar can then be reinforced with a narrow strip of the liner material and joined to the membrane with adhesive (C75).

As previously mentioned, plastic and synthetic rubber membranes should be covered with suitable earth or gravel material as soon as

FIGURE 82. Unfolding and placing polyethylene (PE) plastic lining from a 30 m roll (USBR).

possible to prevent early deterioration by mechanical damage (particularly wind damage) and sun exposure. To prevent dimensional changes due to temperature in hot climates, the covering should be placed at night to avoid damage to the film, which becomes very soft at high temperatures (Figure 83).

Cost

Material costs of plastics and synthetic rubber vary with thickness, width and type. The light weight per unit area permits shipment of these lining materials over long distances at relatively low cost.

It is expected that new materials and manufacturing techniques will reduce costs. Things to consider besides price when selecting materials are availability, cost of delivery to the site, size of sections of material and ease of joining sections, and expected longevity. Since all these characteristics are changing rapidly owing to improved production methods, choice should be based on the latest information available.

FIGURE 83. A dragline carefully dumping earth on a vinyl liner, Nebraska, U.S.A.

Examples

Kirkuk Irrigation Project, Iraq. Figures 84a–d show some stages in the construction of a buried PVC lining in a main feeder canal of this project. The canal is 6.5 m deep and 67 m wide and has a capacity of 300 m^3/sec and a total length of about 32 km. The general type of lining is a buried PVC membrane 0.36 mm (15 mil) thick.

The PVC was delivered in units of 18.3 × 183 m (60 × 600 ft) and joined together by solvent welding. Before placing the membrane, the subgrade was sterilized with a 60 percent sodium chloride weedkiller. Initially, a cover of three layers with a total thickness of about 40 cm was proposed. The first layer was a silty soil covered by a layer of gravel and then covered by riprap. Drawdown tests carried out on a scaled model showed, however, that this cover was highly unstable. Slippage occurred at very low drawdown rates. When silty sand was mixed with the gravel, the stability increased considerably. Gravel alone was not considered practicable as it might damage the thin membrane. Therefore, the membrane was covered by two earth layers which, together, were 30 to 40 cm thick. The lower layer was composed of a silty gravel, and the upper layer of riprap armouring. This was required to meet erosive forces from wave action and stream flow. The design velocity is 1.3 m/sec. The canal side slopes are 4.5:1, which is very flat for a lined canal.

FIGURE 84a, b. Installing PVC-membrane lining on a main feeder canal, Kirkuk Irrigation Project, Iraq.

(a) Section of membrane in place.

(b) Joining the membrane.

FIGURE 84c, d. Installing PVC-membrane lining on a main feeder canal, Kirkuk Irrigation Project, Iraq.

(c) Covering the membrane with a mixture of silty sand and gravel.

(d) Completed section covered with layer of riprap.

The cover materials were hauled and dumped by modern earthmoving equipment, but the spreading of materials on the perimeter of the canal was carried out manually. Figure 85 shows damage caused by overtopping of flood water.

Lining programme, Pakistan. In 1973, the Government of the Punjab, Pakistan, embarked upon a programme of lining more than 70 water courses — totalling more than 320 km — per year in Punjab Province. This province has about 44 000 unlined water courses, in which an estimated average of 11 percent of head discharge is lost through seepage, for a total loss of 7 000 million m^3 of water annually. Lining work is undertaken during six rotational canal closures per year.

Two types of linings were chosen for the programme: (*a*) buried polyethylene (PE); (*b*) brick-cum-polyethylene (PE).

The first type, which is about one quarter as expensive as the second, has been used, in sheets 0.2 mm thick and approximately 3 m wide, wherever local soils are suitable as cover material. A cross section of this type of lining is shown in Figure 86. Profile walls of brick are provided at 30 m spacing. They serve as guides when placing and compacting the cover material and will facilitate future canal clearing. The alternative brick-cum-polyethylene lining is shown in Figure 87.

FIGURE 85. Damage to a covered PVC-membrane lining caused by flood water overtopping the levee, Kirkuk Irrigation Project, Iraq.

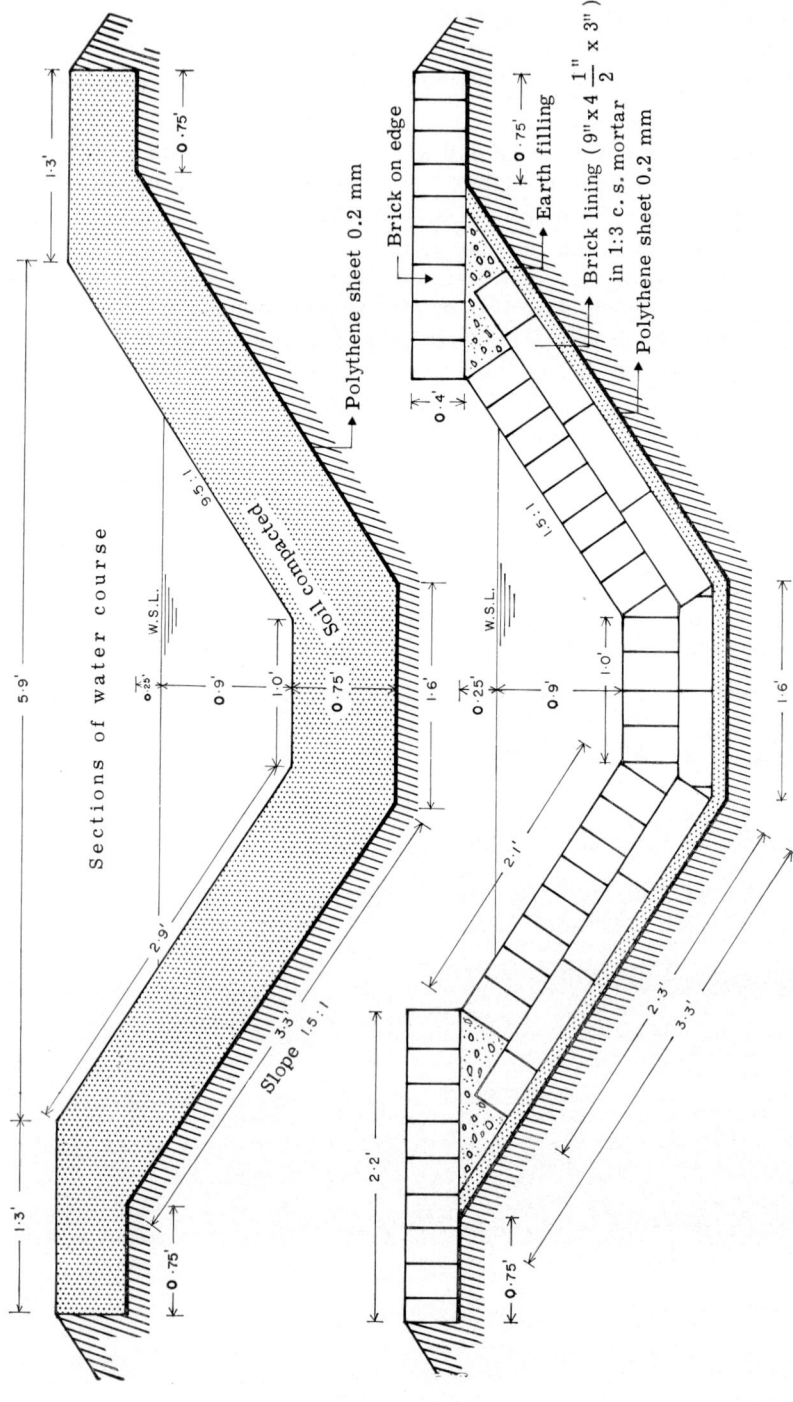

FIGURE 86. *Above*: Water course lined with a buried polythene sheet. *Below*: Masonry profile wall (D10).

NOTE: Polythene is polyethylene used as a plastic.

Section of water course

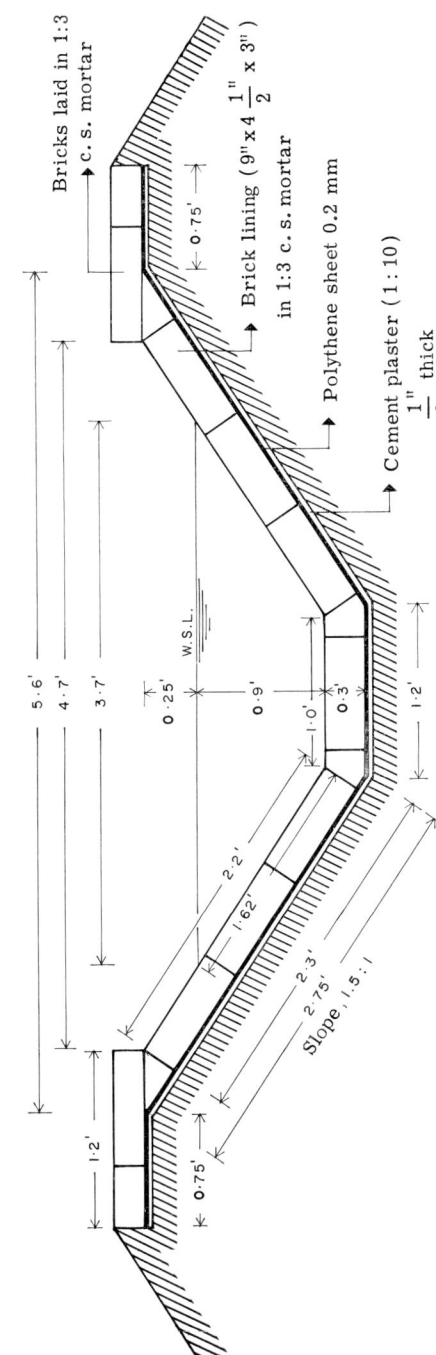

FIGURE 87. Section of water course lined with polythene sheet overlaid by brick masonry (D10). Feet and inches.

Fabric-cum-cement plaster lining

A combined fabric-cum-cement plaster lining developed by a manufacturer in the United Kingdom is now undergoing field testing. The lining consists of a 1.75 mm synthetic rubber membrane with a random-weave polyester mat on the outside surface. This fabric serves as an integrating link between the rubber as the impermeable membrane and the cement plaster as the protecting layer. The cement plaster is pneumatically applied, similar to shotcrete (gunite), and is only 4 to 6 mm thick.

Bentonite membrane linings

Bentonite is a natural clay-like substance formed by the alteration through weathering of volcanic ash. Its chief constituent is the clay mineral montmorillonite, accompanied by small percentages of various other minerals. Bentonite is distinguished from other clays by its extreme fineness, highly absorbent nature and curious property of swelling in water. High-quality bentonite contains up to 90 percent colloidal-size particles.

These clays may be divided into two general classes — high-swelling and low-swelling — depending upon their reaction in water. The high-swelling type, sodium bentonite, absorbs nearly five times its weight of water and at full saturation occupies a volume nearly 15 times its dry bulk; on drying it returns to its original volume. This swelling-drying process is reversible an infinite number of times in pure water, and this capability gives bentonite its water-sealing power (C64). The low-swelling type has a wet bulk 1.5 to 7 times its dry bulk. Although satisfactory results may be obtained by using low-swelling clays in canal linings, more material will be required to secure a desired reduction in seepage than with high-swelling bentonite.

(Testing procedures for the evaluation of bentonite clays are described in C19. They may also be obtained through the U.S. Bureau of Reclamation or through the U.S. Department of Agriculture.)

Bentonite deposits vary greatly in montmorillonite-type clay content, being generally accompanied by sand, silt and clay-sized impurities. Bentonite from various sources therefore differs considerably in expansion characteristics, and care should be exercised in the evaluation of local clays and in relating the membrane thickness to their properties.

While mainly applied for sediment sealing of operating canals and reservoirs, bentonite is sometimes used as a buried membrane. Depending on quality, the bentonite is spread 2.5 to 5 cm thick over the canal subgrade and covered with a 15 to 30 cm protective blanket of stable-earth or gravel.

Bentonite linings have often lost their effectiveness after only a few years of service. A common cause of failure is the decomposing effect of wetting and drying and other weathering factors.

If suitable material is available from local deposits and can be used in its "original" condition, it may be lower in cost than other membrane-type linings. It should be borne in mind that ceramic clays used for making bricks and tiles are not suitable for sealing purposes.

Earth linings

In the course of progress in soil-mechanics engineering and improvement of earth-moving equipment and techniques, the construction of earth linings has advanced so that they have become one of the most common types of irrigation canal lining. In some countries earth linings are surpassed only by concrete linings.

Included in the category of earth linings are those composed of compacted earth, loosely placed earth, clay mixtures, and soils mixed with certain stabilizing additives, such as resins, chemicals, asphalts and petrochemicals. Also straw has been used. The stabilization of soils by using cement and lime is discussed under soil-cement linings (page 108).

GENERAL DESIGN CONSIDERATIONS

A typical section of a compacted-earth-lined canal is shown in Figure 88. The proportions indicated apply to all types of earth linings, except that the lining thickness may vary considerably (see below). If the canal section cuts a pervious as well as an impervious soil stratum, lining of the impervious part is not necessary. Side slopes in earth canals are 1.5:1 or flatter, depending on canal size and materials available for lining, as well as the type of lining to be used. If highly plastic soils are to be used, it is advisable to use slopes 2:1 or flatter because of loss in stability when these soils become saturated.

Freezing and thawing action as well as alternate wet and dry cycles are serious hazards to most types of compacted earth linings. They reduce the soil density of the compacted layers and thus lessen the effectiveness of the lining as a water barrier.

Erosion protection of earth linings is necessary. Linings constructed of silty and sandy materials with little coarse gravel are susceptible to scour. If these are to be used, the cost of reducing the velocity with a larger section, as compared with the cost of maintaining a smaller section with its higher velocity and protecting the lining with a gravel cover,

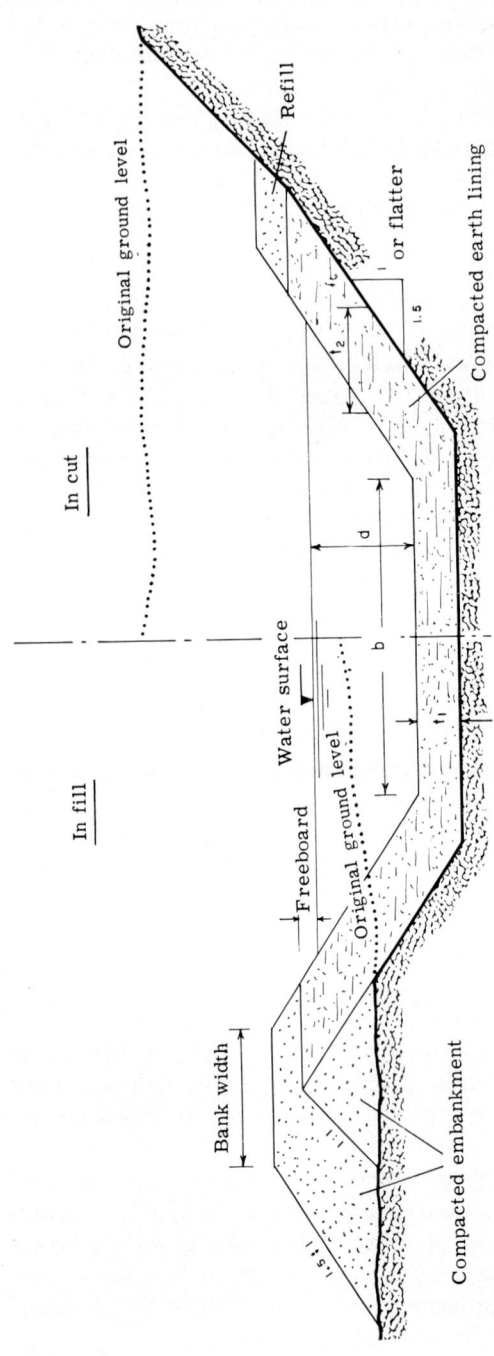

FIGURE 88. Typical section of a compacted-earth-lined canal. (Source: A1 and A3).

Approximate requirements for thick compacted-earth linings

Depth (d)	Bottom thickness (t_1)	Side thickness (t_2)	Bed width to depth (b/d)	Slope (S)
60 cm (2 ft)	30 cm (1 ft)	90 cm (3 ft)	2	1.5:1
120 cm (4 ft)	45 cm (1.5 ft)	120 cm (4 ft)	3	1.5:1 or 1.75:1
250 cm (8 ft)	60 cm (2 ft)	180 cm (6 ft)	3.5	2:1
250 cm (over 8 ft)	60 cm (2 ft)	250 cm (8 ft)	4 to 7	2:1

146

should be evaluated. Tables 17 and 18 will assist in selecting materials for linings and gravel protection if needed.

For the placing of all compacted earth linings, rigid control of the density and optimum moisture is required. A density from 95 to 98 percent of the laboratory maximum, as determined by the Proctor compaction method, will normally provide adequate stability and impermeability. Several in-place density tests should be taken (at least one for each 1 000 m^3 placed) during construction in order to check the quality of the work. If the specific density is not attained, the material should be further compacted. A check on the permeability should also be obtained by laboratory tests on soils taken from the lining. A few field permeability tests should be made with the well permeameter, piezometer, double-tube or other methods (see also Chapter 2). These in-place tests are particularly important at the beginning of each lining job to ensure that the construction procedures are producing the desired results.

The permissible velocities for different earth materials are shown in Tables 3 and 4. A Manning's roughness coefficient of 0.0025 is used for canals with capacities less than 3 m^3/sec (100 cusecs), and 0.020 should be used for larger canals (A1).

THICK COMPACTED EARTH LININGS

Where suitable earth material is available near the site of construction, a lining of thick compacted earth is a cheap and efficient means of controlling seepage. It also can withstand considerable external, hydrostatic uplift pressure without loss of effectiveness, and it can be used, in many instances, without drains under the lining in areas where the canal prism intersects the groundwater table. For similar reasons, a thick compacted earth lining can be used to advantage over expansive clays which produce more rigid types of linings.

A thick compacted earth lining is constructed of selected soils, both the bottom and side slopes being compacted in successive horizontal layers not more than 15 cm thick after compaction (see Figure 88). The thickness normal to the slope may be from 50 to 100 cm depending upon the size of the canal section. Bottom linings are commonly 30 to 60 cm thick.

Suitable soils

The final selection of material for thick compacted-earth linings is made on the basis of several laboratory tests indicating the engineering properties of the soils (gradation, plasticity, compaction and permeability). For identification and classification the Unified Soil Classification System (A1, F3) is commonly used. On the basis of this system, Tables 17 and 18 have been drawn up, showing the properties and suitability of various types of soils for canal linings. The numbers in Table 17 indicate the order of increasing values for the property named. Thus the permeability of the most permeable soil, poorly graded gravel, is indicated as 16, the highest permeability number. The erosion-resistance characteristics of natural sections, cover materials, or earth linings are indicated in the next to the last column in order of suitability, where 1 equals best. Similarly, the last column provides overall information on the suitability of compacted soils for lining purposes in order of suitability.

A well-graded sand and gravel with a clay binder is considered the best material for a compacted earth lining (see Table 17). Clayey gravel soils are considered next best, sand with clay binder the third best, etc. Silty gravels have been used frequently with considerable success even on larger canals. The best materials of this type have just enough silt to provide sealing qualities and have sufficient gravel for erosion resistance. Silty or sandy clay soils having a plasticity index [1] of less than 7 are not considered suitable. Fat clays (i.e. inorganic clays of high plasticity) may not be suitable for canals which are subject to wetting and drying because of swelling and shrinking, unless the lining is protected by a gravel-sand cover.

As the primary function of a compacted earth lining is to control seepage, impermeability of the soil in a compacted state is of prime importance. The water loss of a proposed lining can be estimated from the results of the laboratory permeability test on compacted soil specimens, the thickness of the proposed lining, and the water depth.

After lining, canal losses should not exceed 30 l/m^2 per day (0.1 ft^3/ft^2 per day). If the local or imported soils do not have the desired sealing quality at standard thickness, it will be necessary to increase the lining thickness. In some instances seepage losses as high as 60 l/m^2 per day have been tolerated where the cost of hauling good lining materials was greater than the value of the water saved.

Proper compaction of the lining is essential to increase the stability

---

[1] Plasticity index = The difference in water content as percentage of dry weight at the liquid limit and plastic limit of a soil.

TABLE 17. — IMPORTANT PHYSICAL PROPERTIES OF SOILS AND THEIR USES FOR CANAL LININGS BASED ON THE UNIFIED SOIL CLASSIFICATION SYSTEM (A1)

Major divisions of soils			Typical names of soil groups	Group symbols	Soil properties			Suitability for canals	
					Permeability	Shearing strength	Compacted density	Erosion resistance	Compacted-earth linings
Coarse-grained soils (More than half of material is smaller than No. 200 sieve size) (The No. 200 sieve size is about the smallest particle visible to the naked eye)	Gravels (More than half of coarse fraction is larger than No. 4 sieve size)	Clean gravels (Little or no fines)	Well-graded gravels, gravel-sand mixtures, little or no fines	GW	14	16	15	2	—
			Poorly graded gravels, gravel-sand mixtures, little or no fines	GP	16	14	8	3	—
		Gravels with fines (Appreciable amount of fines)	Silty gravels, poorly graded gravel-sand-silt mixtures	GM	12	10	12	5	6
			Clayey gravels, poorly graded gravel-sand-clay mixtures	GC	6	8	11	4	2
			Gravel with sand-clay binder	GW-GC	8	13	16	1	1
	Sands (More than half of coarse fraction is smaller than No. 4 sieve size) (For visual classifications, the ¼ inch size may be used as equivalent to the No. 4 sieve size)	Clean sands (Little or no fines)	Well-graded sands, gravelly sands, little or no fines	SW	13	15	13	8	—
			Poorly graded sands, gravelly sands, little or no fines	SP	15	11	7	9 coarse	—
		Sands with fines (Appreciable amount of fines)	Silty sands, poorly graded sand-silt mixtures	SM	11	9	10	10 coarse	7 erosion critical
			Clayey sands, poorly graded sand-clay mixtures	SC	5	7	9	7	4
			Sand with clay binder	SW-SC	7	12	14	6	3

149

TABLE 17. — IMPORTANT PHYSICAL PROPERTIES OF SOILS AND THEIR USES FOR CANAL LININGS BASED ON THE UNIFIED SOIL CLASSIFICATION SYSTEM (A1) (*concluded*)

Major divisions of soils			Typical names of soil groups	Group symbols	Soil properties			Suitability for canals	
					Permeability	Shearing strength	Compacted density	Erosion resistance	Compacted-earth linings
Fine-grained soils (More than half of material is larger than No. 200 sieve size) (The No. 200 sieve size is about the smallest particle visible to the naked eye)	Silts and clays	Liquid limit less than 50	Inorganic silts and very fine sands, rock flour, silty or clayey fine sands with slight plasticity	ML	10	5	5	—	8 erosion critical
			Inorganic clays of low to medium plasticity, gravelly clays, sandy clays, silty clays, lean clays	CL	3	6	6	11	5
			Organic silts and organic silt-clays of low plasticity	OL	4	2	3	—	9 erosion critical
	Silts and clays	Liquid limit greater than 50	Inorganic silt, micaceous or diatomaceous fine sandy or silty soils, elastic silts	MH	9	3	2	—	—
			Inorganic clays of high plasticity, fat clays	CH	1	4	4	12	10 volume change critical
			Organic clays of medium to high plasticity	OH	2	1	1	—	—
Highly organic soils			Peat and other highly organic soils	Pt	—				—

TABLE 18. — AVERAGE PROPERTIES OF SOILS

Soil classification group	Proctor compaction		Void ratio, e_o	Permeability, k	Compressibility		Shearing strength		
	Maximum dry density	Optimum water content			@ 20 psi [1] (1.4 kg/cm²)	@ 50 psi [1] (3.5 kg/cm²)	C_o psi	C_{sat} psi	Tan $\varnothing$
	Lb/ft^3	Percent		$Ft/year$	.... Percent				
GW	>119	<13.3	([2])	27 000 ± 13 000	<1.4	([2])	([2])	([2])	>0.79
GP	>110	<12.4	([2])	64 000 ± 34 000	<0.8	([2])	([2])	([2])	>0.74
GM	>114	<14.5	([2])	>0.3	<1.2	<3.0	([2])	([2])	>0.67
GC	>115	<14.7	([2])	>0.3	<1.2	<2.4	([2])	([2])	>0.60
SW	119±5	13.3±2.5	0.37±[2]	([2])	1.4±[2]	([2])	5.7±0.6	([2])	0.79±0.02
SP	110±2	12.4±1.0	0.50±0.03	>15.0	0.8±0.3	([2])	3.3±0.9	([2])	0.74±0.02
SM	114±1	14.5±0.4	0.48±0.02	7.5±4.8	1.2±0.1	3.0±0.4	7.4±0.9	2.9±1.0	0.67±0.02
SM-SC	119±1	12.8±0.5	0.41±0.02	0.8±0.6	1.4±0.3	2.9±1.0	7.3±3.1	2.1±0.8	0.66±0.07
SC	115±1	14.7±0.4	0.48±0.01	0.3±0.2	1.2±0.2	2.4±0.5	10.9±2.2	1.6±0.9	0.60±0.07
ML	103±1	19.2±0.7	0.63±0.02	0.59±0.23	1.5±0.2	2.6±0.3	9.7±1.5	1.3±[2]	0.62±0.04
ML-CL	109±2	16.8±0.7	0.54±0.03	0.13±0.07	1.0±0.2	2.2±0.0	9.2±2.4	3.2±[2]	0.62±0.06
CL	108±1	17.3±0.3	0.56±0.01	0.08±0.03	1.4±0.2	2.6±0.4	12.6±1.5	1.9±0.3	0.54±0.04
OL	([2])	([2])	([2])	([2])	([2])	([2])	([2])	([2])	([2])
MH	82±4	36.3±3.2	1.15±0.12	0.16±0.10	2.0±1.2	3.8±0.8	10.5±4.3	2.9±1.3	0.47±0.05
CH	94±2	25.5±1.2	0.80±0.04	0.05±0.05	2.6±1.3	3.9±1.5	14.9±4.9	1.6±0.86	0.35±0.09
OH	([2])	([2])	([2])	([2])	([2])	([2])	([2])	([2])	([2])

Note: The ± entry indicates 90 percent confidence limits of the average value. [1] The values are based on tests made on samples compacted to Proctor maximum dry density at optimum water content (F4). — [2] Insufficient data.

and frost resistance and to decrease erosion and seepage losses. Some borderline soils which are pervious at natural density become less pervious when compacted. Thus, when some subgrade soils require lining, the required excavation may be suitable for lining when compacted.

It has been found economical to mix two or three soils to obtain the type of material desired. Where gravelly soils are encountered in the canal excavation, silt or clay soils may be imported for mixing with the gravelly soils. From 20 to 50 percent fine material may be required, depending on the characteristics of the fine and the gravelly soils.

The blending of materials often allows a considerable reduction in haul quantities and thus a saving in cost. The blending operation is accomplished inexpensively with mixing machines, with road mixing by blading, or with harrows.

Construction considerations

To keep costs to a minimum the use of construction schemes and suitable designs permitting the use of existing earth-moving and compaction equipment is of extreme importance. In larger canals with bed widths greater than 9 m (30 ft), this poses no particular problem. The bottom lining can readily be compacted by heavy sheepfoot rollers. Likewise, the slope section is wide enough to accommodate conventional earth-moving equipment and rollers. In medium-size canals with bed widths of 1.20 to 9 m (4 to 30 ft), double- or single-drum rollers can be used on the bottom, and single-drum rollers can be used on the slope lining. Transportation and placing of materials may be done with scrapers or the material may be transported to the site by end or bottom dump trucks and placed and levelled by dozer or road patrol equipment, depending on the width of the working surface. If the specified lining thickness on the canal slopes is too thin to accommodate larger equipment, the slope lining may be overconstructed into the canal prism, after which it is trimmed to the required lines. Sometimes the operating width for the thinner slope linings can be increased sufficiently for larger equipment by sloping the layers toward the canal prism. Thus, instead of placing and compacting the material in horizontal layers, layers on a slope up to about 4:1 may be used (Figure 89).

For small canals with a bed width less than 1.20 m, there is a problem of placing and compacting the small thicknesses with heavy equipment. Sometimes this has been solved by overconstructing the lining or by filling and rolling the overexcavated canal prism as one full section (A1, B30). Later, the final canal prism is excavated, and the materials so excavated are used to build up the upper uncompacted portion of the fill or are hauled to nearby portions of the canal for lining.

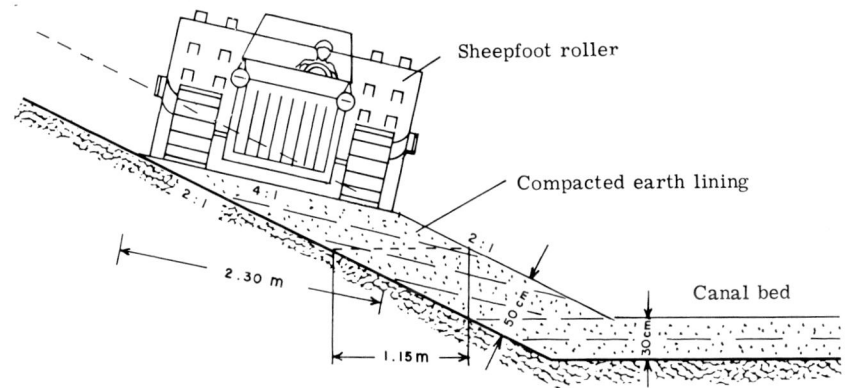

FIGURE 89. Compaction by tractor-drawn sheepfoot roller on a 4:1 slope.

Lovas (C63) reports on the successful use of this construction method for small canals in a pilot project in the Chambal Command Area in India (see Figures 90 and 91). The soil used was a heavy clay, and the main purpose of lining was seepage reduction. The main steps in the construction are described as follows:

1. Topsoil is removed to a depth of 30 to 50 cm.
2. Clay soil is deposited in layers to establish the earthen body of the canal. The thickness of the layer is 50 cm when a 10-ton sheepfoot roller is used for compaction and 15 cm for a 3-ton sheepfoot roller.
3. The soil is compacted to a minimum bulk density of 1.8, for which 4 to 6 passes are necessary.
4. After the embankment reaches the designed height, a cut is made using a " V " bucket machine.
5. The final profile is shaped by manual labour.

Disadvantages are the following:

— Proper compaction requires moisture content of soil near to its plastic limit; if the soil is too wet or too dry, extra measures which increase the cost and time of construction are necessary.
— Deep and dense cracks develop on the surface if the canal is dry.
— A wider strip of land is necessary than for rigid-type linings because of flatter side slopes.

FIGURE 90. Excavation of an irrigation ditch in compacted earth, Chambal Project, India.

FIGURE 91. Compacted earth canal after nine months in operation, Chambal Project, India.

Despite these disadvantages the method is considered more economical than other types of lining in this project area.

In projects where heavy earth-moving and compaction equipment is unavailable or uneconomical, but where a surplus of cheap labour is available, the placing and compacting of earth linings in small canals pose no particular construction problem. The use of simple implements such as box mallets and of animal-drawn carts and rollers for spreading and compacting earth in small canals or short reaches may be as effective as the more modern mechanical equipment described above.

THIN COMPACTED EARTH LININGS

This type of lining may be considered where

— highly suitable earth material is available, providing relative watertightness with thicknesses of only 15 to 30 cm (6 to 12 inches);
— suitable earth material has to be hauled considerable distances, thus necessitating its most economical use;
— the canal can be operated so as to avoid more frequent wetting and drying of the lining;
— lower velocities can be economically permitted to prevent any scour on the thin lining, or coarse soil or gravel is available near the site for a protective cover.

The construction of thin compacted earth linings is best accomplished with a dragline on the canal berm and compaction by transverse rolling from the berm.

COMPACTION OF SLOPES AND BED OF UNLINED EARTH CANALS AND DITCHES

This construction procedure usually consists of scarifying, adding moisture, and compacting the subgrade to the required density by sheepfoot rollers, flat rollers or other available equipment. A similar technique of decreasing the permeability by compacting the canal subgrade has been used in the U.S.S.R. Dadayev (A3) states that there are advantages to compacting the natural soils in place, without overexcavating and reworking, with the objective of utilizing the natural bond between soil particles to achieve a soil structure of greater strength. He also states that in dry areas this method may be more economical because moisture control is not necessary. Seepage tests on a canal with a capacity of 50 m^3/sec before and two years after construction showed a reduction in seepage of 96 percent with this method.

For compacting soils of natural texture, machines with impact, vibration or vibro-impact action must necessarily be used. A number of machines have been developed for this purpose in the U.S.S.R. (A3).

Sharov (E7) states, "Packing of the ground by rolling and tamping is the simplest way of reducing porosity and permeability of soils. With a ten percent reduction of pore spaces, permeability becomes 5 to 7 times less in clay. Permeability of sandy soils is made 3 to 4 times less by the same degree of packing. In order to render packing more effective and to destroy, at the same time, cavities produced by animals and roots, the ground should be first loosened by ploughing a 15-20 cm layer and slightly wetting it before tamping."

LOOSELY PLACED EARTH LININGS

This type of lining consists essentially of a loose, uncompacted earth blanket of selected clay soils dumped into the canal and spread over the bottom and banks to approximate line and grade in layers up to about 30 cm in thickness. Seepage can often be reduced to an acceptable amount economically, provided available soils are sufficiently fine to be impervious in a loose state and are adequately stable to resist erosion to a reasonable degree.

Although the serviceable life of loosely placed earth linings is relatively short, their use may be advantageous in certain cases where low first costs are essential. Costs are low because simple equipment can be used and very little trimming and shaping of the subgrade are necessary.

On one project, consideration was being given to placing a 15 cm blanket of loose plastic clay secured from " borrow " over pervious silty sand-gravel subgrade soils. It was found that the seepage rate of the loose clay blanket was 300 $1/m^2$ per day (1.0 ft^3/ft^2 per day), while the seepage rate of a 60 cm compacted lining of materials obtained from excavation was only a quarter of this seepage rate. Therefore, the latter soils were used for a thick compacted lining.

SOIL MODIFICATION

If available soils are below standard requirements for slope stability, erosion resistance and watertightness, their natural deficiencies can be overcome by treatment with small quantities (1 to 6 percent) of certain substances. These include asphalts, chemicals, lime, petrochemicals, cement, specially treated resins, swelling clays and combinations of these materials. The treatment is restricted to a layer of up to 30 cm within the compacted earth lining (or near its sides).

Treatment with cement is practically identical to the use of soil-cement (see page 108). Little information is available on the effectiveness and economy of the other materials mentioned when used for canal lining. Some experimental installations in the U.S.A. using asphalt proved satisfactory shortly after construction, but no information is available on their condition after some years of service.

COST OF EARTH LININGS

The most important factors influencing the unit cost of earth linings are the following.

Size of the job. A job involving the placement of large quantities of lining in large canals permits the effective use of heavy equipment, thus reducing the cost per square metre considerably. However, price may be relatively independent of job size in projects where manual labour is predominant.

Size of canal. Usually, the larger the canal, the lower the unit cost when standard earth-moving machines are used.

Quality of material. If the local material is unacceptable and must be improved, the cost will increase when there is need for mixing with hauled soils, increasing lining thickness or providing a protective cover.

Source of material. The least expensive linings are those for which materials removed in the required canal excavation can be used in the lining. The economics of importing soil will vary according to each separate set of lining conditions.

Others factors affecting the cost of earth linings are weather conditions such as freezing and optimum natural moisture for compaction, subgrade preparation, canal maintenance and operation such as wetting-drying actions.

The cost for compacted earth linings varies considerably, according to the prevailing conditions on a project. For some lining projects in Canada it was found that thick compacted clay linings with gravel cover cost as much as gravel-covered plastic membrane linings. For larger canals the cost of compacted earth linings was competitive with that of other lining materials (B30).

Wineland and Lucas (B4), having made comparative cost estimates for 140 km of the San Luis Division Canal of the California Acqueduct with a capacity varying from 370 to 230 m^3/sec, concluded: "When comparing concrete linings with heavy compacted earth linings (3 feet thick), a 5-inch concrete lining cannot be economically justified. Considering the accuracy of the analysis, a 4-inch concrete lining may be used in lieu of a heavy compacted earth lining."

Soil sealants

Soil sealants are natural or artificially processed materials which can be injected into flowing or standing water, sprayed in place or injected in subsurfaces to reduce seepage losses in canals and reservoirs.

A natural sealing of operating canals occurs if the water in the canal carries considerable silt- or clay-sized sediments. The sediments penetrating the voids in the subgrade material will gradually clog the soil voids, thus reducing permeability. This phenomenon has been occasionally utilized by artificially adding sealants of various kinds to the flowing or ponded water. Materials used are natural silts and clays, bentonite, resinous polymers, petroleum-based emulsions, cationic asphalt emulsions, sodium chloride, sodium carbonate or soda ash (Na_2Co_3), sodium pyrophosphate ($Na_4P_2O_7$), and others (see *Covered membrane linings*, page 129). Amount and method of application vary widely. Specific information is available from relevant industries and from some research laboratories.

The lining resulting from treatment with soil sealants is usually a relatively thin layer which may drastically reduce seepage losses. However, this layer is highly susceptible to erosion, puncture, wetting-drying actions and destruction by cleaning operations.

Spraying the sealant on the subgrade and mixing it into the soil is a treatment similar to soil modification. Sealants which have been used are a blend of cationic asphalt emulsion, white gasoline and water, sodium carbonate and sodium pyrophosphate. For treatment with these sodium salts the subgrade soil should contain at least 15 percent clay (C71).

On some experimental pond-sealing installations (C81), the subgrade was disked to a depth of 15 cm and supplied with about 1.2 kg of sodium carbonate per square metre (about 5 tons/acre) by a fertilizer or lime spreader, then disked again. When using sodium pyrophosphate the application of only 0.5 kg/m^2 was found equally successful. The performance of the treatment was encouraging from the first to the third year for those parts of the treated area permanently covered with water.

Experience with soil sealants shows that they usually provide a good sealing action during the first few years of service but then rapidly deteriorate. Reduction in water losses of 65 to 90 percent have been recorded a short time after treatment, but no continuing effectiveness is achieved unless the treatment is periodically repeated. As the cost of soil-sealing treatments is low compared with most other lining methods, repeated applications of sealants may be economically justified. Apart from this, the use of soil sealants may also be an economic means of saving water in unlined canals during exceptional water shortages.

Flumes, pipes and lay-flat tubing

FLUMES

The term flume is used to designate an artificial water channel of wood, metal, concrete or masonry that is usually supported above the surface of the ground (Figures 92 to 95).

Entire flume networks are commonly found in countries around the Mediterranean basin and other areas with similar conditions — hilly topography, limited water resources occurring in quite small streams, high returns per unit of water, quite well-developed infrastructure with intensive use of arable land and local availability of concrete materials.

Flume networks must be based on proper engineering designs, and installation requires skilled workmen. The investment cost per unit of water-carrying capacity is quite high compared with lined canal systems.

FIGURE 92. Wooden flume for water distribution to furrows on sugarcane land, Hawaii.

FIGURE 93. Precast concrete flume with siphon under a farm road, Japan.

FIGURE 94. Concrete flume, Spain.

FIGURE 95. Flume of sandstone (Katla) slabs and bricks, Rajasthan, India.

The mechanization of concrete precasting has promoted the installation of concrete flume-type irrigation distribution systems in certain countries, mostly in the Mediterranean region. Since almost every manufacturer has developed his own system, it is beyond the scope of this publication to enter into design and construction details. Furthermore, the development of new types and new manufacturing and construction techniques makes it necessary to base any economic study and design of such systems on information from current commercial sources.

PIPELINES

Although pipe is usually not classified as a lining, it should be considered an alternative to lined canals for conveyance and distribution of irrigation water (Figure 96). The use of irrigation pipe is rapidly increasing in countries with highly developed irrigated agriculture. Materials used are concrete, steel, aluminium, asbestos cement, fibreglass, plastics and polyester resins. The most widespread types are reinforced and nonreinforced concrete pipe. These may be precast or cast in place. Precast pipe is usually manufactured in short lengths at a central plant, transported to the job, laid in a trench, caulked at the joints, and back-

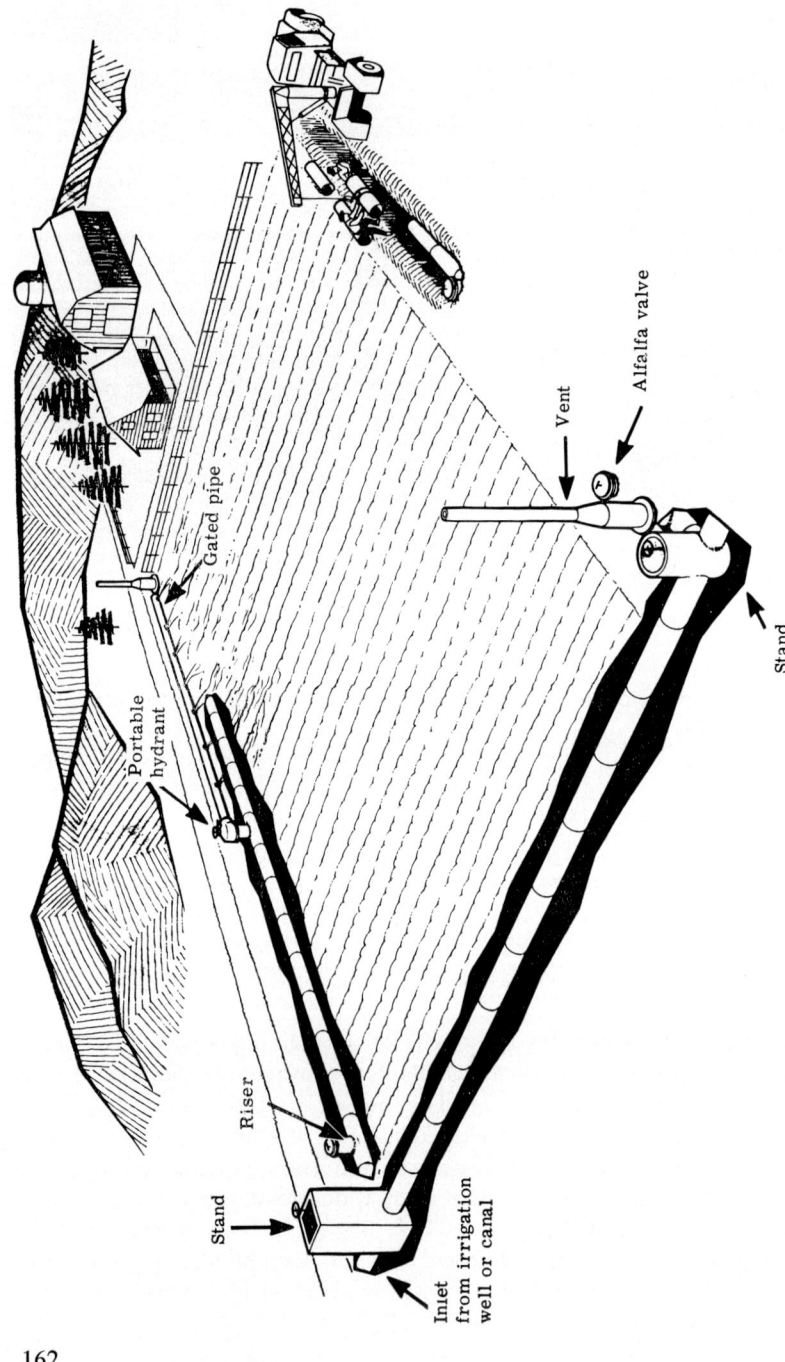

FIGURE 96. Drawing of a buried-pipe delivery system.

filled. Cast-in-place concrete pipe is nonreinforced and constructed in its final position in the trench. (For design details, see A1, A7, C28, C31, C37, C68 and C90.)

High-pressure pipe-conveyance networks are particularly suited for sprinkler and trickle irrigation. Low-pressure pipe is installed for surface irrigation. The main advantages of pipe are listed below:

1. Little loss of land: Practically all of the system is buried, while open canals often take 3 to 5 percent of the land area out of cultivation.
2. Labour-saving operation and maintenance.
3. Adaptability to topography: Pipe-irrigation systems operate under water pressure. Pressure pipe can be laid uphill and downgrade permitting the irrigation of land too rough for open canals. This may reduce the total length of the network considerably, as compared with open canals. Pressure pipes do not have to be installed on a uniform grade as is required for open canals.
4. Close control of water distribution: Greater accuracy in measuring water flow.
5. Evaporation and seepage are minimal.
6. Weed seed contamination is reduced or eliminated.
7. Severances are eliminated.

One disadvantage of pipe systems has been the problem of detecting and sealing leaks. However, in recent years a very promising sealing method, using anhydrous ammonia, has been developed in the U.S.A. and standard field procedures are being developed (C25).

The main limitations of pipe are its high initial cost and the requirement of skilled workmen for installing the pipe. As a rule of thumb, the investment costs for installing precast, nonreinforced concrete pipe are comparable with those of concrete-lined ditches for capacities less than 85 to 140 l/sec (3 to 5 cusecs). For larger flows, the costs of cast-in-place concrete pipes are comparable with those of concrete-lined canals for capacities up to 850 l/sec (30 cusecs). The use of pipe should be considered in the following situations:

— in urban and other areas with intensive cultivation of high-value crops;
— where right-of-way costs are high and where full use of irrigable land is important;

FIGURE 97. Gated tubing to irrigate a tomato crop.

— where the value of water is very high owing to scarcity and/or high delivery cost, necessitating efficient use;

— where the water saved can be utilized for additional high-quality crop production.

In 1973 a new type of so-called ductile cast-iron pipe up to 1.6 m in diameter was installed for irrigation water conveyance in the Federal Republic of Germany. The pipe is said to be superior to other types in strength, corrosion resistance and resistance to mechanical action (*Journal Wasser und Boden*, No. 2, 1974).

Another recent development comes from Japan. There, a firm has developed a PVC pipe which is manufactured in a soft state and can be coiled or folded for shipment. The pipe is hardened by a heat process at the construction site and, while warm, is rounded by air pressure. The pipe obtains its design stiffness in a matter of minutes. At present, pipes up to 1 m in diameter are being field tested in the U.S.A. (W.A. Lidster, Bureau of Reclamation, Regional Representative, Denver, Colo.).

LAY-FLAT IRRIGATION TUBING

This is a relatively recent development, representing an intermediate between flume and pipe discussed previously. It is usually made from

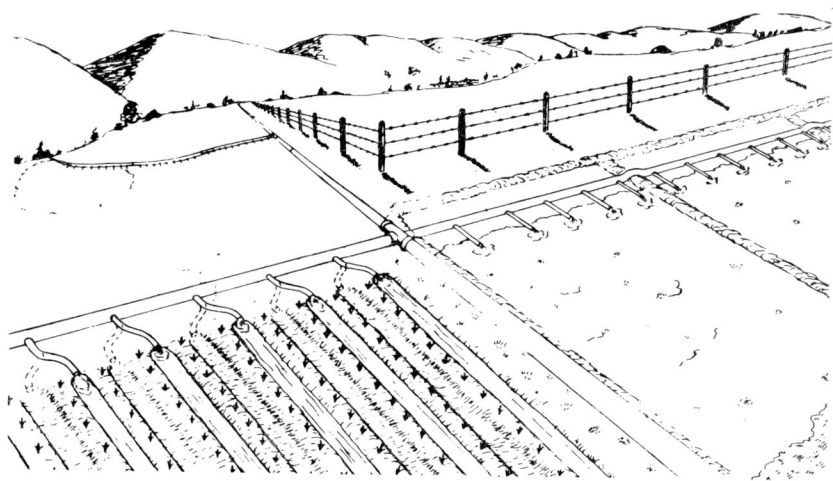
FIGURE 98. Some methods of irrigating crops with lay-flat tubing.

such materials as butyl rubber or plastic in which a supporting fabric is often built into the walls to impart additional strength. It lies flat when empty like a deflated inner-tube section.

The tubing may be used for water distribution as a temporary water conveyance or as a conveyance where a canal would be undesirable or impractical.

Synthetic rubber tubing is now manufactured in 15 m (50 ft) lengths that one man can carry and connect in the field. The individual lengths are provided with ring-type couplings, which provide a watertight joint that is easily connected and disconnected. The tubing is made in two types: transmission and gated. The transmission tubing currently comes in six sizes, from 10 to 40 cm (4 to 16 inches) in diameter with pressure head rating from 12 to 6 m (40 to 20 ft) of water respectively. The 40 cm tubing will transmit as much as 0.34 m^3/sec (12 cusecs) at rated head. Its smooth interior allows high velocities from a low head pump or a reservoir.

The gated tubing consists of short lengths of small-diameter tubing attached to transmission tubing at regular intervals for diverting the water on to the land as shown in Figures 97 and 98.

The practice is to have enough tubing so as to be able to leave it in place. A good mechanical moving device could change this. To permit tillage and other operations the irrigator uncouples and pulls back the tubing as required.

4. SELECTION OF TYPE OF LINING

Although canal linings are very simple structures from an engineering point of view, the fact that normally large investments of labour and materials are involved necessitates very careful selection and design of the lining to be used. Lining needs attention from the very beginning of project planning. There still are advocates of the "wait and see" procedure — that is, of constructing and operating unlined canals and judging whether to line or not according to the findings, so that lining is avoided if seepage losses are negligible. Other advantages may be earlier project completion and the possibility of future employment of rural labour for lining works during off-seasons. However, this system does not make use of the possible economic advantages of lining at the time of canal construction. These advantages may include:

— Less right-of-way cost and better utilization of land due to smaller canal cross section.

— For a given area the possibility of canal structures, diversion and storage facilities and pumping installations of smaller capacity.

— Shorter bridges and other crossings.

— Appropriate location and gradient of canal and the avoidance of certain drop structures.

— Reduced excavation, except for earth linings.

— Saving in drainage investment for prevention of waterlogging.

— No interruption of irrigation water supply.

— Saving of additional cost for by-passes during lining of existing canal.

Water losses through seepage and its secondary effects are normally the governing factors in lining decisions. In order to avoid unjustified expenses for lining reaches in which seepage is negligible, water losses along a proposed canal can be predicted by several means (see Chapter 2).

This prediction helps determine the need for lining as discussed in Chapter 1. In close connection with this procedure, the type of lining which best suits given conditions must be decided. The purpose of this chapter is to outline the criteria for lining selection and to draw up some main guidelines for an optimal solution. The more common types of lining and their main features are summarized in Table 19.

Factors governing the selection of lining

The sequence in which these factors discussed below does not necessarily reflect the order of importance, as this depends on local conditions.

Soil properties

Failures have occurred where concrete and some other rigid-type linings have been constructed on subgrades containing swelling clays or gypsum. Some failures have also been experienced on cavernous limestone rock. Where canals have to be excavated in such soils or formations, a more flexible type of lining, such as thick compacted earth lining or a buried membrane lining, would serve better. Sometimes, when short reaches of unsuitable soils have to be crossed, it may be advantageous to remove the unsuitable soil to a certain depth by overexcavation and replace it with sand or other suitable material as subgrade for a hard-surface lining. In some cases, it may be possible to bypass areas with unsuitable subsoils by changing the canal alignment. Soil surveys may serve as a guide for canal location.

It is important to examine the soils along or near the line of a proposed or existing canal for possible use as lining materials.

If sufficient amounts of sand and gravel are contained in the canal excavation or are to be found within reasonable hauling distances, they may favour the choice of a concrete or soil-cement lining. If the soils show suitability for compaction, a compacted earth lining should be seriously taken into consideration. If the excavated soil can be used neither as aggregate nor for compaction, it may still be useful as cover material for a membrane lining, although great attention has to be paid to its erosion and to resistance to drawdown (piping hazard). Improvement by mixing with borrow-pit soils may also be considered.

If the canal embankments as well as the natural in-place soils have to be replaced or compacted to serve as a subgrade for a rigid-type lining, a compacted earth lining or another flexible-type lining is likely to be more economical.

TABLE 19. — IRRIGATION CANAL LININGS AND THEIR MAIN FEATURES

Type of lining and thickness	Durability (service life)	Water losses (m³/m²/24 h)	Other important features
A. Hard-surface linings			
Portland cement concrete, unreinforced, 5 cm	Commonly estimated to last 50 years	Below 0.03 if well constructed and maintained, but values up to 0.15 have been measured	Suitable for all sizes of canals, all topographical, climatical and operational conditions; firm subsoil required; susceptible to swelling clays; availability of aggregates near the job is essential; construction either by hand methods or slipform.
As above, but 7.6 cm			
As above, but 10 cm and reinforced			
Pneumatically applied mortar, unreinforced, 5 cm	In mild climate and stable subgrade same as concrete (30 years have been reported)	0.03-0.06	As above, but no need for coarse aggregates; special equipment necessary; generally not economical for large jobs; suitable on subgrades of weathered rock.
Precast concrete blocks, 7 cm	About the same as above if properly maintained	If joints are well sealed, about 0.03 can be achieved	Advantageous where concrete lining is suitable, but remote precasting is more economical (lack of aggregates at site, transport facilities for precast material available).
Soil-cement (dry-mix), 13 cm	Largely dependent on cement content;	0.03-0.06	Although less durable than portland cement concrete, low initial

Soil-cement (plastic), 7.6 cm	23 years have been recorded		costs make this an economic lining where suitable sandy soils are available from canal excavation or nearby.
Asphaltic concrete, in place, 5 cm	Seldom more than 15 to 20 years	About 0.03, but will increase considerably if weed infested	For the in-place type, availability of aggregates at site is essential; because of shorter service life, asphaltic concrete does not offer any advantage over cement concrete except on less stable subgrades (swelling clays); offers better resistance against certain chemical deterioration; susceptible to weed penetration.
Asphaltic concrete, prefabricated slabs, 3.8 cm			
Brick and stone	May be as high as cement concrete if properly constructed and maintained	Brick with cement plaster: around 0.03. Stone: relatively permeable unless carefully mortared	Labour-intensive methods; availability of construction material at or near the site is essential.
B. *Exposed membranes*			
Asphaltic materials	Only a few irrigation seasons	Vary widely depending on weed penetration and other mechanical damage as well as weathering	Suitable only as temporary lining for seepage control.
Polyvinyl (0.19 mm; 8 mil)			
Resins			
Synthetic rubber (1.44 mm; 60 mil)	Not yet known, but not less than 10 years	Negligible if properly jointed and maintained	Offers permanent seepage control if protected from physical damage but is high in cost.

TABLE 19. — IRRIGATION CANAL LININGS AND THEIR MAIN FEATURES (*continued*)

Type of lining and thickness	Durability (service life)	Water losses ($m^3/m^2/24$ h)	Other important features
C. *Buried membranes*			Suitability of excavated soil as cover material is important for economic reasons.
Sprayed-in-place asphalt	Depends largely on erosion resistance of cover material, maintenance (weed hazard, beaching, burrowing animals), and operation (drawdown); records show a serviceable life of at least 15 years, but rubber membrane is likely to last much longer	Below 0.06	Heater and spray equipment must move along canal; skilled personnel are required.
Prefabricated asphaltic membrane		Below 0.08	Easily transported and placed materials, but slippage of cover material caused particularly by drawdowns has sometimes been a problem.
Polyethylene (0.24 mm; 10 mil)		Below 0.06	
Polyvinyl (0.24 mm; 10 mil)		As above	
Synthetic rubber (0.77 mm; 32 mil)		Below 0.03	
Bentonite layer (4-5 cm)	Not reported	—	—
Bentonite layer (1.3 cm)	Less than 7 years	—	After 7 years, water losses equal to unlined conditions.

Sublining of plastic sheeting or sprayed-in-place asphalt under precast concrete	Determined by service life of concrete lining	Practically watertight if properly constructed	Very effective in preventing seepage. Concrete joints and cracks need not be sealed but eventually filled with some material to protect the underlying membrane.
D. Earth Linings			
Thick compacted (approx. 90 cm thick) Thin compacted (30 cm and less)	For economy evaluations 20 years have been assumed	Below 0.08 (0.02 has been measured)	Suitable soil from canal excavation or nearby borrow pit area is essential for economy. Freezing-thawing and alternate wetting-drying are hazards to all compacted-earth linings because they loosen the compaction and increase the permeability.
Loosely placed earth (loam, clay)	—	—	Low initial cost, but with little effectiveness as to seepage control; little advantage against unlined canals; low durability.
E. Soil sealants			
Waterbone bentonite Sodium carbonate Resinous polymers, petroleum, asphalt emulsions and other chemicals sprayed on the subgrade	One or two irrigation seasons	May average around 0.30 after treatment but varies widely	Means of temporarily controlling seepage in unlined canals. Sealing effect is high just after treatment but may be reduced to less than half after only one or two irrigation seasons. Because of low cost, repeated treatment may be an economical alternative to more durable types of lining.

TABLE 19. — IRRIGATION CANAL LININGS AND THEIR MAIN FEATURES (*concluded*)

Type of lining and thickness	Durability (service life)	Water losses (m³/m²/24 h)	Other important features
F. *Flumes and pipes*			
Concrete flumes	Approx. 50 years	Negligible if joints are well sealed	Relatively independent of soil and topographic conditions; ratio of cost to carrying capacity is high; economical only when value of water is very high.
Concrete pipes (precast, cast in place)	More than 50 years	Negligible if joints are properly sealed	Particularly suitable for areas with irregular or rolling topography and intensive cultivation.
Lay-flat tubing	Not yet known	Practically nil	As above.

Topography

A high-pressure pipe system is the most independent means of conveying and distributing irrigation water with regard to topography, but its use is limited by its high cost. Reasonably adaptable to terrain conditions are low-pressure pipes and flumes, followed by concrete-, brick- and stone-lined canals. Earth linings and buried membranes are normally best suited for slightly sloping or flat land, mainly because of limited permissible velocities, which may be one sixth those in concrete canals. Furthermore, hard-surface-lined canals can better follow contour lines because the radius of curves chosen can be much smaller than in earth-lined canals.

Fewer elevation-reducing structures are needed, because of the possible steeper gradient of hard-surface-lined canals, and where structures are needed they can be constructed cheaper than on earth-lined canals, where larger stilling basins and scour protection measures may be required.

If a canal is to be located on a hillside, smaller canal width, by employing steeper slopes, and safety against slippage are important considerations normally leading to the adoption of a concrete lining.

Water table

If the water table is above the bed level of a canal, emptying the canal can be expected to cause external hydrostatic pressure on the lining. Unless drainage facilities are provided, this pressure has to be met by the dead weight or elasticity of the lining. Failures have been reported where linings were too light and too rigid to resist the pressure. This is true especially for thin bituminous linings, cement mortars and linings of thin concrete, brick or stone slabs. Heavy compacted earth linings have always proved satisfactory in this respect.

In colder climates concrete linings are not recommended in areas with a high water table, as frost expansion will crack the lining when the canal is empty.

Land use and irrigation systems

In urban and other areas of high land value and intensive cultivation, the fullest possible use of irrigable land is made and right-of-way costs are normally high. Such conditions favour the installation of buried pipe systems, flumes or hard-surface-lined canals with steep side slopes. It would be uneconomical in such cases to employ, for example, a buried membrane lining which normally requires 3:1 side slopes.

This problem is of particular relevance when improvements of traditional irrigation systems and patterns are to be undertaken, such as land consolidation, replacement of continuous flow delivery by rotation delivery, and crop diversification. Such modifications usually result in an increase in total length of ditches and higher capacity requirements for certain existing canals and ditches. In both cases hard-surface linings to minimize loss of land to the distribution system should be considered. They at the same time meet the need for better water control in a rotation delivery system.

OPERATION AND MAINTENANCE

If the operation of a canal system requires frequent filling and emptying or causes frequent water level changes, a hard-surface lining will perform best. With earth linings, including earth-covered membranes, such conditions would speed up the deterioration process considerably and would require increased maintenance efforts. With regard to weed control, small repairs and silt removal, such linings offer little advantage over unlined canals. Where labour is expensive, the higher maintenance cost of earth or membrane in comparison with most hard-surface linings may even exceed the difference of installation cost.

The adoption of exposed membrane linings, thin clay or thin compacted earth linings may be limited by the hazard of livestock traffic, cleaning operations, or vandalism.

When existing canals are to be lined, the time available for executing the work may influence the choice of lining. If an interruption of flow or an off-season shutdown for several months is possible, any method of installation may be chosen. If interruptions of only a few weeks are possible, mechanized placing methods may be preferable.

The off-season shutdown during which lining can take place may be too short to allow the invert of the existing canal to dry out sufficiently for proper installation of a hard-surface lining. Here, overexcavation and installation of a compacted earth lining or covered membrane may be the better solution if the canal is large enough.

In small canals and ditches the economy of compacted earth linings and covered membranes becomes critical. One reason is the unfavourable ratio of overexcavation to total excavation in small cross sections. Another reason is that such linings require considerable maintenance, are more difficult to secure and more often neglected on small water courses than on large. Employing a high-quality hard-surface lining cannot however compensate for disregard of maintenance.

For the permanent lining of canals during operation, no economical method has been found so far. A temporary sealing effect can be achieved by applying waterborne soil sealants. A method using grouted fabric mats is mentioned in Chapter 3.

WATERTIGHTNESS

If the value of water is high and the measurement or estimation of seepage shows high losses, obviously the aim should be to adopt a relatively watertight lining. Probably the most impermeable and lasting lining is a thin plastic, asphalt or rubber membrane placed under a normal concrete lining. The additional cost for such a "sublining" of some membrane material has been justified on several lining projects because of the additional water saved. On one project in the U.S.A. the seepage for a 10 cm concrete lining was estimated to average 21 l/m^2 per day (0.07 ft^3/ft^2 per day), which a PVC sublining would reduce by 95 percent. From this and the value of the water saved, it was found that employing the sublining would pay if it would remain efficient for at least 12 years (B4).

DURABILITY

The durability of linings depends on type of lining, quality of the construction materials used, quality and accuracy of installation, climatic conditions, canal operation and canal maintenance. Properly constructed and maintained cement-concrete linings should normally have a service life of at least 40 years. In many countries a rate of 50 years is applied for this type in economic evaluation, and some mortar linings in small canals in California have served for more than 60 years.

Joseph (D6) reports that with normal repairs carried out annually the life of a cement-brick-lined canal is expected to be 20 years. Similarly, he takes the life of buried concrete pipelines as 50 years, with normal maintenance at 0.1 percent. For a plaster of asphalt-clay mix, as used in India, he attributes a service life of 5 years, with 10 percent per year for maintenance. He estimates the service life of a simple mudplaster lining 2.5 cm thick as no more than two years, with a maintenance cost of 25 percent.

The durability assigned to a lining will essentially influence the benefit-cost calculation and should therefore be determined carefully. Performance of lined canals in nearby projects may provide a valuable indication as to the life expectancy. In general, the higher the figure of durability chosen for a certain type of lining, the higher the yearly allowance for maintenance of that lining in the economic evaluation.

The life expectancy of waterborne soil sealants and thin exposed membranes may be shorter than two years, but may still be economically feasible as temporary linings under certain conditions. If water losses in a canal are high but in normal years do not affect water demands, a permanent lining may not be justified. However, to avoid a shortage in drought periods, a cheap temporary lining may be feasible.

Compaction of natural soils in unlined canals (see Chapter 3) is another temporary method for preventing seepage. Its effectiveness may last for only one irrigation season; however, by repeating the treatment during the off-season a satisfactory level of seepage reduction may be maintained. Specific data on durability are found in Table 19 and in Chapter 3.

AVAILABILITY OF CONSTRUCTION MATERIALS

For all heavy permanent-type linings it is essential, for economic reasons, that the material used to construct the lining be available at the site or within reasonable distances from the canal. Traditional linings such sons, that the material used to construct the lining be available at the site

Natural or selected soils are undoubtedly the best materials for canal lining and need careful examination. The extent of utilizing off-site material depends largely on transport facilities. If both cement and aggregates have to be transported long distances to the job, serious consideration should be given to prefabrication, or a lining other than concrete should be employed. In sandy soils dry-mix soil-cement requiring only the procurement of cement may be a cheap and suitable lining.

If the canal water contains considerable amounts of silt, the self-sealing effect should be taken into account and could perhaps make artificial lining unnecessary.

Generally the most economical lining is that which makes the best use of locally available materials.

AVAILABILITY OF LABOUR AND MACHINERY

Some linings are suitable for manual labour and others, like concrete, for machine installation. The choice, therefore, is often governed by the relative supply of labour and machinery. In countries and regions with a surplus of manual labour, it is a sociopolitical necessity to adopt methods which utilize the labour potential to the fullest extent and at the same time conserve foreign exchange. Linings of tile or concrete bricks, as commonly found in some Asian countries, are typical examples of highly labour-intensive methods.

Even compacted earth linings in small to medium-sized canals may be economically carried out by manual labour aided by simple implements and animal power. However, when the thickness of the compacted earth exceeds one-half metre, as in large canals, mechanical earth-moving and compaction equipment such as bulldozers, scrapers, and sheepfoot rollers are required. If this equipment cannot be made available at a reasonable cost, and manual and animal labour are the main power sources, it would be more economical to utilize this power to construct a brick lining than to aim at a thick compacted earth lining.

Often, prompt completion of an irrigation project is necessary in order to obtain early benefits. This calls for the employment of rapid lining methods, which may include the use of machine rather than manual techniques.

If a certain method of canal lining is established in the project area or region because it has proved satisfactory and beneficial, its continuation should be considered, however, with the aim of improving traditional methods whenever possible. To adopt a pattern from established practices may have several advantages: local enterprises and workmen are familiar with the execution; means of processing and purchasing materials and construction equipment may be available; local operation and maintenance practices can be adopted; comparative technical and economic data may be readily available for more accurate project planning.

COST AND FINANCIAL ASPECTS

Costs of a given lining have to be weighed against obtainable benefits. Theoretically the most economical solution should be adopted regardless of costs; however, in practice the financial resources of a project determine whether lining, and which type, can be afforded.

Where construction costs for the most economic but expensive solution exceed available funds, a choice must be made between

— adopting a less expensive lining, which allows for the initial lining of the whole network within the given financial limit; or

— constructing the more expensive lining at the initial stage of canal construction as far as funds permit, leaving the rest unlined until further funds permit completion. (This is called staging.)

Since lining should coincide with initial canal construction if full use of its various advantages is to be taken, it is usually more economical to line a large portion of the network with a less expensive lining rather than to line only a small portion with a costlier type. Under the following conditions, however, the second choice may be the best solution:

— if initial lining does not offer major advantages, or if a temporarily unlined network can be constructed to fit the final lined condition as regards grade, size, side slope, structures;
— if off-seasons are sufficiently long to proceed with the lining work in stages;
— if there is underemployment during off-seasons or available maintenance forces.

In addition, a shortage of foreign exchange may limit the use of certain lining materials and equipment.

Benefit-cost analysis

By evaluating the physical and economical conditions required for the lining decision, keeping in mind the previously discussed factors, the planner will soon direct himself toward the more promising lining conceptions. The final selection among competing solutions should be based on their individual benefit-cost ratios. Obviously, these ratios must be positive if lining is to be feasible at all. For the first approach it may be sufficient to include only the main or tangible benefits. If they do not lead to a distinct solution, those benefits which are more difficult to determine or intangible should be added to justify a particular lining. They should be estimated rather conservatively, because there is a danger of overestimating.

Cost should be calculated on an annual basis in order to reflect the specific service life and current cost of the different linings. Initially an asphalt concrete lining may cost half as much as a concrete lining, but it may have only half the service life of the latter and require more maintenance, thereby proving more expensive in annual cost than the concrete lining.

The amount of water saved will usually be the determining factor in the cost-benefit confrontation. A realistic estimate will consider that losses in a lined canal usually increase with years of service and that losses in an unlined canal will usually decrease because of the natural sealing effect (Figures 2 and 3).

Cost figures for lining should not be "borrowed" from other projects or from the literature, but should be calculated for the specific project conditions.

The following example illustrates how to calculate and compare annual costs and benefits. The example (derived from A7) refers to an unlined canal and a canal lined with portland cement concrete. The unit costs are illustrative only. (Similar examples are to be found in A4, B4, D6 and D7.)

Assumptions

1. Required canal capacity: 28.3 m^3/sec.
2. Loose sandy soil, level, economic cut permissible (see glossary: *Economic cut*).
3. Permissible velocity, unlined: 0.50 m/sec.
4. Permissible velocity, lined: 1.50 m/sec.
5. Maximum water depth: 3.10 m.
6. Bank slopes, unlined: 2:1.
7. Bank slopes, lined: 1.5:1.
8. Portland cement concrete lining, unreinforced: 7.6 cm (3 in) thick.
9. Top width of banks: 4.25 m.
10. Shrinkage allowance: 10 percent.
11. Seepage loss, unlined canal: 0.46 m^3/m^2/day (1.5 ft^3/ft^2/day).
12. Seepage loss, lined canal: 0.015 m^3/m^2/day (0.05 ft^3/ft^2/day).
13. Manning's *n*, unlined canal: 0.0225.
14. Manning's *n*, lined canal: 0.013.
15. Available slope: 0.0002.
16. Cost of excavation, including placing in embankments: $0.65 per m^3.
17. Cost of lining: $4.30 per m^2.
18. Annual maintenance cost, unlined canal: $1.15 per m.
19. Annual maintenance cost, lined canal: $0.50 per m.
20. Value of water: 0.33 cent per m^3.
21. Life of lining: 40 years.
22. Interest rate: 5 percent per annum.

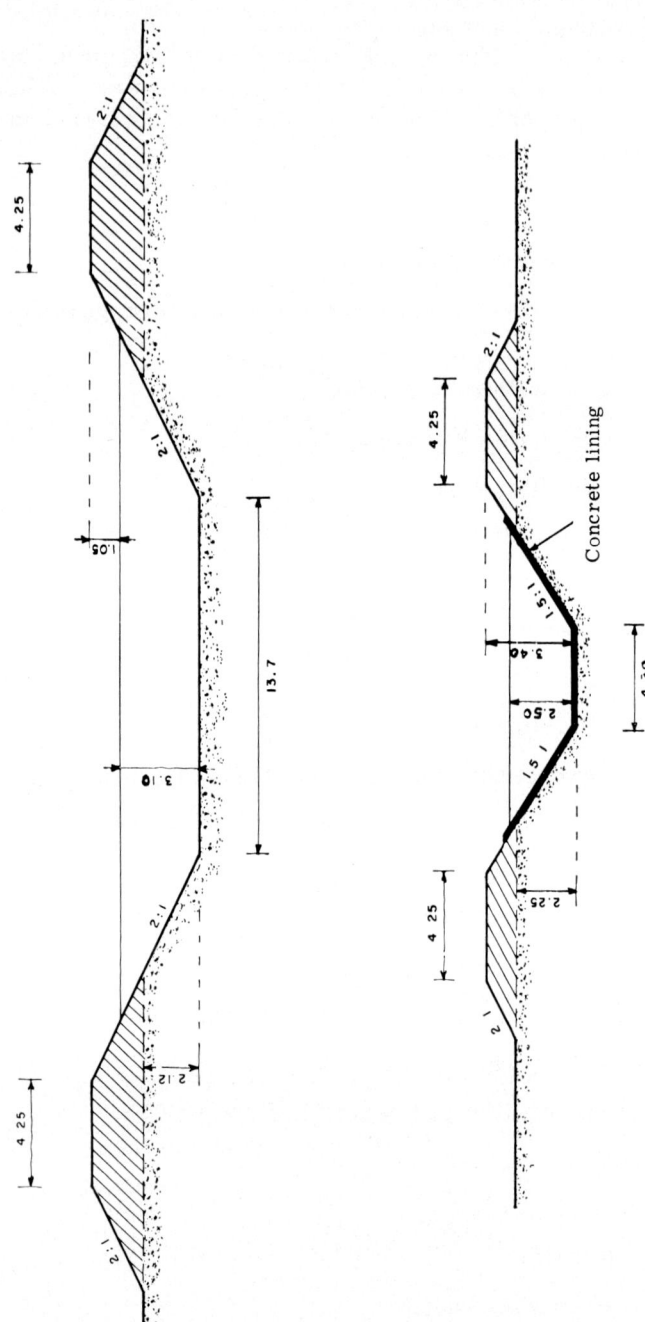

FIGURE 99. Comparative cross sections of an unlined and a concrete-lined canal (in metres).

180

23. Variable operation over 8-month period equivalent to about 6 months' continuous full flow.

24. Land area lost to the unlined canal: 47 m² per metre of canal.

25. Land area lost to the lined canal: 31 m² per metre of canal.

26. Annual right-of-way costs: $180 per ha.

In this example the annual right-of-way costs are considered equal to the farmer's net income per hectare of land. It is further assumed that the land produces 3 tons of rice per hectare a year, which gives the farmer a gross income of $300 per hectare. Assuming 40 percent for inputs, excluding the farmer's own labour input, the net return is $180 per hectare per year.

For an unlined canal, the permissible velocity of 0.50 m/sec calls for a cross-sectional area of 62 m², which is satisfied by a 13.7 m bottom width and a depth of 3.10 m, with r = 7.36 and s = 0.000036. This is a very flat slope, less than one third of that available. The remainder must be consumed in falls, the cost of which is not considered here. The resultant section, using a freeboard of 1.05 m, is shown in Figure 99. The economic cut is 2.12 m. The excavation is 38.0 m³/m.

The seepage loss on a wetted perimeter of 27.5 m is 12.6 m³/m of canal per day. This is equivalent to 2 300 m³/m of canal for the specified operating season.

For the lined canal the allowable depth of 3.10 m calls for a bottom width of only about 2 m, which is too narrow for efficient construction. If a base width of 4.25 m is chosen, the velocity is 1.43 m/sec and the water depth is 2.50 m. The total depth is 3.40 m when a freeboard of 0.90 m is added. Allowing for the lining thickness, the total depth is about 3.50 m and the excavation bottom width is 4.30 m. The economic cut is 2.25 m and the volume of the excavated materials is 17.3 m³/m. If the concrete lining is carried 0.60 m above the water surface, the lined perimeter is 15.4 m. The wetted perimeter is 13.2 m. The leakage at 0.015 m³/m² per day is 0.20 m³/m of canal per day, or 36 m³/m of canal per operating season.

Annual cost per metre of unlined canal

(*a*) Capital cost:

excavation: 38 m³ at $0.65 = $24.7

interest: $= \dfrac{24.7 \times 0.05}{2} =$ $0.62

(b) Value of annually lost water: 2 300 m³ at 0.33 cent per m³ = $7.60
(c) Maintenance = $1.15
(d) Right-of-way cost: $\dfrac{180 \times 47}{10\,000}$ = $0.85

Total annual cost = $10.22

Annual cost per metre of lined canal

(a) Capital cost:
    excavation: 17.3 m³ at $0.65 = $11.25
    concrete lining: 15.4 m² at $4.30 = $66.22

    $77.47

    interest: $\dfrac{77.47 \times 0.05}{2}$ = $1.65

    depreciation of lining: $\dfrac{66 \times 22}{40}$ = $1.93

(b) Value of annually lost water: 36 m³ at 0.33 cent per m³ = $0.12
(c) Maintenance = $0.50
(d) Right-of-way cost: $\dfrac{180 \times 31}{10\,000}$ = $0.56

Total annual cost = $4.76

Annual saving per metre of lined canal: $10.22 minus $4.76 = $5.46

On the basis of the assumed data, the lined canal saves $5.46 per metre per year. The annual saving would be zero if the water losses in the unlined canal became equal to or less than 0.13 m³/m² per day (0.43 ft³/ft² per day), which corresponds to a clay or clay loam (see Table 2). In this case other benefits, such as savings on canal structures and increased safety, may still justify the concrete lining.

Assuming the same seepage rate as in the above example, but varying the water value, the adoption of the lining would still be justified if the water value were to drop to approximately 0.10 cent per cubic metre.

5. SUGGESTIONS FOR FURTHER RESEARCH AND DEVELOPMENT

1. The development of more precise criteria for choosing between concrete or other hard-surface-lined open-channel networks and buried pipe networks is important.

2. There is very little information available on the changes in effectiveness of lining with the passage of time. Such knowledge would greatly improve predictions of the economic feasibility of lining investments. It therefore appears important to investigate by means of case studies the benefits actually obtained by investments in canal lining.

3. World practices show considerable disagreement on quality, thickness and reinforcement of cement concrete linings. Guiding criteria based on field experience in different parts of the world would be beneficial.

4. Much of the seepage loss in unlined or earth-lined canals is attributable to a too low natural soil density or a deterioration of initial density due to weathering and other effects (wetting-drying, freezing, piping, burrowing animals, vegetation, etc.). Travelling machinery for compaction of the invert of earth canals and ditches has been tested in the U.S.S.R. (A3). Elsewhere, this method does not appear to have been given the attention it deserves. It is believed that investment in the development of efficient, robust machinery for use by maintenance forces to compact earth canals and ditches as a normal recurrent maintenance procedure in off-seasons should be promoted as a possible alternative to lining. Such machinery should have vibro-compaction devices which can be adjusted to fit a certain range of canal sections and depths. A combination of pressure and vibration should be exerted on the canal perimeter while the machine proceeds along the canal (either pulled or self-propelled). The treatment may be done in several passes and may comprise the entire perimeter of small canals and ditches.

Epilogue

An ever-increasing number of irrigation canals and ditches are being lined in order to save valuable water and land, to minimize deterioration of the environment through seepage, and to obtain other benefits. The need to make water conveyance more efficient or to satisfy certain conveyance criteria appears to have been well met by technology; there is hardly any material, natural or artificial, which has not been tried as canal lining; the installation techniques are manifold, and there is even a marked trend away from open toward closed conveyance networks. The different methods all have their specific merits and limitations; also, there are many situations in which lining is not required at all. The amount of water lost in earth canals remains the principal yardstick for the decision to line canals (or to use covered pipe). Water losses should not be arbitrarily estimated and should be distinguished from operational losses and from leakage through structures and other losses not recoverable by lining canals. The wide choice, ranging from no lining at all to reinforced concrete lining installed to last for half a century or more, together with the large investments usually involved in lining work, makes careful assessment and planning essential.

GLOSSARY

AGGREGATE. The inert mineral materials, such as sand, stone dust, gravel, shells, slag, broken stone, or combinations thereof, with which cement, lime or bituminous material is mixed to bind into a mortar or concrete.

AIR-ENTRAINED CONCRETE. Concrete produced with the addition of purposeful air-entraining agents to improve durability and other properties.

ALIGNMENT. The course in plan along which the centre line of a canal or drain is located.

ASPHALT. A mix of bitumen and mineral aggregate. In the United States the term asphalt refers to the product known in the United Kingdom and most other countries as bitumen.

BERM, OR INSIDE BERM. (*a*) Horizontal strip of shelf built into an embankment or cut to break the continuity of an otherwise long slope, usually for the purpose of reducing erosion or to increase the thickness or width of cross section of an embankment; (*b*) space left between the upper edge of a cut and the toe of an embankment; (*c*) nearby horizontal formation along a beach caused by the deposit of material under the influence of waves.

BORROW PIT. A source from which material is borrowed to complete a section and to make fills.

BULK-DENSITY. The ratio of the mass of water-free soil to its bulk volume. Bulk-density is expressed in grams per cubic centimetre. Equal to apparent specific gravity or volume weight. Also APPARENT DENSITY.

CANAL. Artificial open water channel for the conveyance or distribution of water. The term "canal" in this manual generally comprehends main, secondary, tertiary channel, water course, open conduit, distributary, branch, feeder, lateral and so on.

Cohesion material, or coherent material. Material possessing sufficient shear strength to resist the shearing stresses of the flow of water and therefore not readily scoured.

Concrete. A suitably proportioned mixture of aggregates (including sand) with cement and water, or lime and water, or bituminous material. When cement or lime is the binding agent, the plastic mass of concrete which can be cast or moulded into a predetermined size and shape hardens by hydraulic chemical reaction. The concrete with cement or lime is called "cement concrete" or "lime concrete" respectively; concrete in which bitumen and aggregates may be hot- or cold-mixed is called bituminous or asphalt concrete.

Curing, or maturing. The process of keeping concrete or mortar damp for the first week or two of its life to prevent or replenish the loss of necessary moisture during the early, relatively rapid stage of hydration. Sometimes also referred to as "maturing."

Depreciation. The loss in service value of a project due to (*a*) wear and tear not covered by current repairs; (*b*) obsolescence or inadequacy resulting from age, physical change, supersession by reason of inventions, discoveries, change in public demand or public requirements; (*c*) destruction of property by extraordinary casualties.

Ditch. Artificial open water channel for distribution of irrigation water at farm or field level. Normally the capacity of ditches is less than 0.5 m^3 per second (about 15 ft^3 per second).

Durability. Resistance of lining to weathering, chemical attack and wearing.

Economic cut. For practical purposes there exists within a one-bank canal section a point on the ground surface which, if the excavation is to balance the embankment, will have the same cut and will be the same distance from the canal centre line for all ground surface slopes so long as they are uniform. In other words, for balanced conditions, all cross-slope lines pass through this point. The cut at this point is called the "economic cut," the point is called the "pivot point," and the distance from it to the centre line is called the "offset distance."

FLOATING. Smoothing the surface of newly placed concrete or mortar with a trowel.

FREEBOARD. The difference between the maximum flow line and the top of the bank of a canal.

HOLIDAY. An area, or spot, inadvertently missed in the coverage by asphalt.

HYDRAULIC CONDUCTIVITY. The proportionality factor in the Darcy flow law which states that the effective flow velocity is proportional to the hydraulic gradient. Hydraulic conductivity, therefore, is the effective flow velocity at unit hydraulic gradient and has the dimensions of velocity.

INVERT. The bottom or floor of an irrigation canal or lateral.

MAINTENANCE. The operations performed in preserving irrigation canals in good or near-original condition without increasing capital costs. Repairs are part of maintenance.

MOST ECONOMIC SECTION. Cross section which with a given surface slope and given amount of excavation has the maximum capacity.

PERMEABILITY. See HYDRAULIC CONDUCTIVITY.

PERMISSIBLE VELOCITY. The highest velocity at which water may be carried safely in a canal or other conduit. The highest velocity throughout a substantial length of a conduit that will not scour.

PIPING. The process of formation of wash-ins.

PORTLAND CEMENT. The product obtained by finely pulverizing clinker produced by calcining to incipient fusion, an intimate and properly proportioned mixture of argillaceous and calcareous materials, with no additions subsequent to calcination except water and calcined or uncalcined gypsum.

PUDDLING. The process by which a soil loses granular structure and becomes deflocculated. It is caused by excessive water, excessive handling or tilling or deflocculating agents.

Rubble. Loose water-worn stones; rough, irregular fragments of broken rock.

Screed. Mechanically operated strike-off unit designed to the proper crown or surface cross section, which removes excess material ahead of it as it progresses across a surface.

Slope, steepness. Expressed in this manual as a ratio, with the horizontal distance given first and followed by the vertical distance which is always one (for example, 5:1, 1.5:1).

Slump test. A standard British test for consistency of concrete. An open-ended mould of frustum cone shape is placed on a levelled flat plate and filled with concrete rammed in a specified manner. The top surface is struck off and the mould carefully lifted clear. The amount by which the concrete subsides below the top of the mould is measured and is defined as the "slump."

Steady flow. A constant flow; that is, the same volume in equal units of time. Also steady-state flow, permanent flow.

Unsteady flow. Flow in which features such as velocity, cross-sectional area and hydraulic slope vary in the course of time.

Waterlogging. State of land in which the subsoil water table is located at or near the surface with the result that the yield of crops commonly grown on it is reduced well below the normal for the land.

Weathering. The geological processes, caused by physical and chemical action of atmospheric agencies upon rock at or near the surface of the lithosphere, which result in the more or less complete disintegration and decomposition of such rock, and in some instances, its removal to other locations by separate wind and water action. Also, the action of atmospheric agencies in altering the colour composition or formation of any substance.

BIBLIOGRAPHY

A. *General*

1. U.S. BUREAU OF RECLAMATION. *Linings for irrigation canals.* Washington, 1963 D.C. 149 p.

2. PORTLAND CEMENT ASSOCIATION. *Lining irrigation canals.* Chicago, Ill. 42 p. 1949

3. *Canal lining.* International Commission on Irrigation and Drainage, Third 1957 Congress. Question 7.

4. PORTLAND CEMENT ASSOCIATION. *Lining irrigation canals.* Chicago, Ill. 33 p. 1957

5. INTERNATIONAL COMMISSION ON IRRIGATION AND DRAINAGE. *Controlling seepage* 1968 *losses from irrigation canals: worldwide survey 1967.* New Delhi. 100 p.

6. AHMED, N., AHMED, M. & SHAK, S.L. *Problems of canal lining in West* 1957 *Pakistan.* International Commission on Irrigation and Drainage, Third Congress, Question 7. 30 p.

7. DAVIS, C.V. *Handbook of applied hydraulics,* p. 415-440. 2nd ed. New York, 1952 McGraw-Hill.

8. HAGOOD, M.A. *Lining canals and reservoirs.* Rome, FAO. 18 p. (Unpublished) 1967

9. HOUK, I.E. *Irrigation engineering.* Vol. 2. *Projects, conduits and structures.* 1956 New York, Wiley. 531 p.

10. ISRAELSEN, O.W. & HANSEN, V.E. *Irrigation principles and practices.* 3rd ed. 1962 Chapter 5. Conveyance of irrigation and drainage water, p. 75-92. New York, Wiley.

11. LAURITZEN, C.W. et al. *Lining canals and reservoirs to reduce conveyance* 1952 *losses.* Logan, Utah Agricultural Experiment Station. Circular No. 129.

12. LAURITZEN, C.W. & TERRELL, P.W. Reducing water losses in conveyance and 1967 storage. *In* American Society of Agronomy, *Irrigation of agricultural lands,* p. 1105-1119. Wisconsin.

13. OSMAN EL NUR, MUSTAFA. *Report on canal lining.* Rome, FAO. 4 p.

14. SANFORD, H. *Canal lining. Scope of the subject: purpose and practical* 1957 *aspects, particularly materials, technical and economical aspects,*

maintenance and related problems. Review of the papers presented. International Commission on Irrigation and Drainage, Third Congress. Question 7. 28 p.

15. ZIMMERMAN, J.D. *Irrigation.* New York, Wiley. 516 p.
    1966

16. INTERNATIONAL COMMISSION ON IRRIGATION AND DRAINAGE. *Design practices*
    1972   *of irrigation canals in the world.* New Delhi. 276 p.

B. *Seepage losses*

1. FAO.   *Report to the Government of Turkey on water management studies in*
   1965   *the Cumra irrigation area.* Rome. FAO/EPTA Report No. 1975. 228 p.

2. FAO.   *Report to the Government of Turkey on irrigation problems of the*
   1963   *Konya-Cumra Plain and in the vicinity of Antalya and Izmir.* Rome. FAO/EPTA Report No. 1664. 12 p.

3. U.S. AGRICULTURAL RESEARCH SERVICE. *Proceedings of the Second Seepage*
   1968   *Symposium.* Phoenix, Ariz. ARS-41-147. 150 p.

4. U.S. AGRICULTURAL RESEARCH SERVICE. *Proceedings of the Seepage Sym-*
   1963   *posium.* Phoenix, Ariz. ARS-41-90. 180 p.

5. Conveyance losses in irrigation canals. *Civil Engineering,* 12: 584-585.
   1941

6. U.S. BUREAU OF RECLAMATION. *Water Measurement Manual.* 2nd ed. Denver,
   1967   Colo.

7. U.S. BUREAU OF RECLAMATION. *Irrigation operators' workshop.* Vol. 2.
   1964   Lecture notes. Denver, Colo.

8. U.S. SOIL CONSERVATION SERVICE. ENGINEERING DIVISION. *Irrigation water*
   1967   *requirements.* Washington, D.C. Technical Release No. 21. 83 p.

9. FAO/Unesco. *Irrigation, drainage and salinity: an international source book.*
   1967   Chapter 10. Irrigation systems and management. [London], Hutchinson/FAO/Unesco.

10. BOUWER, H. Theory of seepage from open channels. In *Advances in*
    1969   *hydroscience.* Vol. 5. New York, Academic Press.

11. BOUWER, H. & RICE, R.C. Seepage meters in seepage and recharge studies.
    1963   *Journal of the Irrigation and Drainage Division, American Society of Civil Engineers,* 89: 17-43.

12. BOUWER, H. Rapid field measurement of air entry value and hydraulic
    1966   conductivity of soil as significant parameters in flow system analysis. *Water Resources Research,* 2: 729-738.

13. BOUWER, H. & RICE, R.C. Modified tube diameters for the double-tube
    1967   apparatus. *Proceedings, Soil Science Society of America,* 31: 437-439.

14. BOERSMA, L. Field measurement of hydraulic conductivity above a water
    1965   table. *In* Black, C.A., ed. *Methods of soil analysis.* Madison, Wis., American Society of Agronomy. Agronomy Monograph No. 9.

15. BROCKWAY, C.E. & WORSTELL, R.V. Field evaluation of seepage measurement
    1968   methods. *In* U.S. Agricultural Research Service. *Proceedings of the Second Seepage Symposium*, p. 121-127. Phoenix, Ariz. ARS-41-147.

16. COOKE, F.T. *An economic analysis of factors affecting water loss in irrigation*
    1961   *channels in the Yazoo-Mississippi Delta.* State College, Miss., Mississippi Agricultural Experiment Station. Bulletin 626. 7 p.

17. CORREIA, J.F. A study of the factors affecting seepage and methods for
    1963   evaluating losses due to seepage in small irrigation channels. *Journal of Soil and Water Conservation*, 18: 5-16.

18. DHILLON, G.S. Estimation of seepage losses from lined channels. *Indian*
    1967   *Journal of Power and River Valley Development*, 17: 16-20.

19. DHILLON, G.S. Measurement of seepage losses from irrigation canals. *Indian*
    1967   *Journal of Power and River Valley Development*, 17: 23-28.

20. DHILLON, G.S. Estimation of seepage losses from unlined channels. *Indian*
    1968   *Journal of Power and River Valley Development*, 18: 317-324.

21. DOORENBOS, J. *A literature survey of seepage in canals. Preliminary report.*
    1963   Wageningen, International Institute for Land Reclamation and Improvement.

22. GLOVER, R.E. *Groundwater movement.* Washington, D.C., U.S. Bureau of
    1964   Reclamation. Engineering Monograph No. 31. 67 p.

23. HORST, L. *Notes on seepage loss experiments for Mbarali Irrigation Scheme.*
    1961   *Rufiji Basin Survey.* Rome, FAO. Special Report 61/C/2026. 22 p.

24. KIRKHAM, D. Saturated conductivity as a characterizer of soil for drainage
    1965   design. *Proceedings of the Intersociety Conference on Drainage for Efficient Crop Production, Chicago, Illinois,* p. 24-32. Madison, Wis., American Society of Agricultural Engineering.

25. LINSLEY, R.K. & FRANZINI, J.B. *Water resources engineering*, p. 267-271.
    1964   New York, McGraw-Hill.

26. LOVAS, L. *Report on the investigations of water losses in the main canals*
    1970   *of the Chambal Commanded Area.* FAO/UNDP/SF Project IND 60.

27. LAURITZEN, C.W. Ways to control losses from seepage. *In* U.S. Department
    1955   of Agriculture. *Water: yearbook of agriculture*, p. 311-320. Washington, D.C.

28. NELSON, K.D. *et al.* Distribution losses in irrigation channels. *Civil Engi-*
    1966   *neering Transactions, Institution of Engineers, Australia*, CE8 (2): 179-182.

29. PACHINSKY, A.A. The nature of losses in irrigation canal systems. Measures
    1963   on reduction. In *International Seminar on Main Irrigation Canals and their Equipment, Tashkent,* p. 203-259.

30. POHJAKAS, K. & RAPP, E. *Performance of some canal and dugout linings on*
    1967   *the Canadian prairies.*

31. ROBINSON, A.R. & ROHWER, C. *Measuring seepage from irrigation channels.*
    1959   Washington, D.C., U.S. Agricultural Research Service, Colorado Agricultural Experiment Station and U.S. Department of the Interior. Technical Bulletin 1203.

32. ROHWER, C. & VAN PELT STOUT, O. *Seepage losses from irrigation channels.*
    1948   Fort Collins, Colo., Colorado Agricultural Experiment Station. Technical Bulletin 38.

33. STAROSOLSZKY, O. Measuring irrigation water for investigating the efficiency
    1962   of irrigation systems. *In* International Commission on Irrigation and Drainage. *Annual Bulletin 1962*, p. 36-44.

34. SCOTT, V. & HOUSTON, C. *Measuring irrigation water.* Berkeley, Calif., Cali-
    1959   fornia Agricultural Experiment Station, Extension Service. Circular 473.

35. STAROSOLSZKY, O. Common errors in measurement of irrigation water. Discus-
    1959   sions — C.W. Thomas. *Transactions of the American Society of Civil Engineers*, 124. Paper No. 2980.

36. SOLETAUCHE. *Hydraulique souterraine. Théorie élémentaire de l'hydraulique*
    1964   *des puits et application pratique.* Paris. 36 p. (Unpublished)

37. SOLETAUCHE. *Hydraulique souterraine. Les essais de pompage. Quelques don-*
    1966   *nées théoretiques et pratiques.* Paris. 80 p. (Unpublished)

38. TODD, D.K. *Groundwater hydrology.* New York, Wiley. 336 p.
    1959

39. VAN DER WEERD, B. *Apparatus voor het meten van slootkwel* [Apparatus for
    1966   the measurement of seepage]. Wageningen, International Institute for Land Reclamation and Improvement. Mededelingen 95. 9 p. (In Dutch)

40. VAN DER VEEN, C. *Report to the Government of Pakistan on seepage and*
    1962   *sediment problems in the Kushita Unit of the Ganges-Kobadak Irrigation Scheme.* Rome, FAO. FAO/EPTA Report No. 1519.

41. VAN BEERS, W.F.J. *The auger-hole method: a field measurement of the*
    1963   *hydraulic conductivity of soil below the water table.* Wageningen, International Institute for Land Reclamation and Improvement. Bulletin 1. 32 p.

42. WEUZEL, L.K. *Methods for determining permeability of water-bearing ma-*
    1942   *terials with special reference to discharging well methods.* Washington, D.C., U.S. Geological Survey. Water-Supply Paper 887. 192 p.

43. LUTHIN, J.N. & KIRKHAM, D. A piezometer method for measuring perme-
    1949   ability of soil *in situ* below a water table. *Soil Science*, 68: 349-358.

44. INTERNATIONAL INSTITUTE FOR LAND RECLAMATION AND IMPROVEMENT. *Veld-*
    1972   *boek voor land en water des Kundigen* [Field book for land and water practice]. Wageningen. 672 p. (In Dutch)

C. *Design and construction of lining*

1. INTERNATIONAL INSTITUTE FOR LAND RECLAMATION AND IMPROVEMENT. *The*
   1964   *design of open watercourses and ancillary structures.* Wageningen. Bulletin 7. 80 p.

2. INDIA. CENTRAL BOARD OF IRRIGATION AND DRAINAGE. Design of channels.
   1965   *In Irrigation research in India*, p. 47-57. New Delhi. Publication No. 78.

3. ARIZONA. AGRICULTURAL EXPERIMENT STATION. *Suggested construction methods*
1951 *and specifications. Concrete lined farm irrigation ditches.* Tucson, Ariz. Report No. 106. 10 p.

4. GUNITE CONTRACTORS' ASSOCIATION. *Water control with Gunite.* Los Angeles, Calif. Bulletin No. 129. 4 p.

5. ASPHALT INSTITUTE. *Asphaltic concrete. Asphalt canal lining.* New York. Construction Series No. 87. 3 p.

6. ASPHALT INSTITUTE. *Buried membrane asphalt canal lining.* New York. Construction Series No. 89. 2 p.

7. U.S. BUREAU OF RECLAMATION. *An evaluation of jute-reinforced prefabricated*
1963 *asphaltic canal lining after two and three years' field service.* Washington, D.C. Lower Cost Canal Lining Programme. Report No. B-31A. 15 p.

8. ASPHALT INSTITUTE. *Prefabricated asphalt canal lining.* New York. Construction Series No. 89. 3 p.

9. U.S. BUREAU OF RECLAMATION. *Evaluation of field ageing on the physical*
1964 *characteristics of buried hot-applied asphaltic membrane canal lining.* Washington, D.C. Lower Cost Canal Lining Programme. Report No. B-34. 21 p.

10. U.S. BUREAU OF RECLAMATION. *Tentative specifications: canal lining, plastic film, polyvinyl chloride.* Washington, D.C. 7 p.

11. U.S. BUREAU OF RECLAMATION. *Evaluation of cationic asphalt emulsion as a*
1963 *waterborne canal sealant by hydraulic flame testing.* Washington, D.C. Lower Cost Canal Lining Programme. General Report No. 32. 18 p.

12. U.S. BUREAU OF RECLAMATION. *Isolation and evaluation of canal subsurface*
1963 *membranes produced by a waterborne petroleum-based, emulsion-type sealant. Westside lateral, Eden project, Wyoming.* Washington, D.C. Lower Cost Canal Lining Programme. Report No. B-33. 21 p.

13. *Handbook of instructions for the design of lined and unlined channels and masonry works.* Bhakra-Nangal Project, India.

14. U.S. BUREAU OF RECLAMATION. *A review of the use of chemical sealants for*
1960 *reduction of canal seepage losses.* Washington, D.C. Lower Cost Canal Lining Programme. Analytical Laboratory Report No. CH-102. 14 p.

15. FAO. *Report to the Government of Turkey on water management studies in*
1965 *the Cumra irrigation area.* Rome. FAO/EPTA Report No. 1975. 228 p.

16. FAO. *Report to the Government of the United Arab Republic on lower cost*
1959 *canal linings for Egypt.* Rome. FAO/EPTA Report No. 1062. 18 p.

17. FAO. *Report to the Government of Pakistan on estimates and designs for*
1961 *canal systems and installation of automatic regulators in the Kushita Project in East Pakistan.* Rome. FAO/EPTA Report No. 1399. 18 p.

18. U.S. BUREAU OF RECLAMATION. *Drawdown tests on earth cover material*
1961 *placed over an asphalt membrane.* Washington, D.C. East Bench

Canal-Missouri River Basin Project, Montana. General Report No. GEN-29. 12 p.

19. COLORADO AGRICULTURAL EXPERIMENT STATION. CIVIL ENGINEERING SECTION. 1965 *Evaluation of Colorado clays for sealing purposes.* Fort Collins, Colo. Technical Bulletin 83. 35 p.

20. U.S. BUREAU OF RECLAMATION. *Irrigation operators' workshop.* Vol. 1. 1965 Lecture notes. Denver, Colo.

21. PORTLAND CEMENT ASSOCIATION. *Soil-cement laboratory handbook.* Chicago, 1959 Ill. Code No. SC 6-5.

22. PORTLAND CEMENT ASSOCIATION. *Soil-cement for paving slopes and lining ditches.* Chicago, Ill. Code No. SCB 14-2.

23. PORTLAND CEMENT ASSOCIATION. *Cost estimate form for soil-cement construction.* Chicago, Ill. Code No. SCB 15.

24. PORTLAND CEMENT ASSOCIATION. *Soil-cement construction handbook.* Chicago, Ill. Code No. EB 0003.08.

25. U.S. BUREAU OF RECLAMATION. *Annual report of progress on engineering* 1968 *research.* Washington, D.C.

26. U.S. BUREAU OF RECLAMATION. *Laboratory and field investigations of plastic* 1968 *films as canal lining materials.* Washington, D.C. Open and Closed Conduits Systems Programme. Report No. CHE-82.

27. SHELL INTERNATIONAL PETROLEUM COMPANY LTD. *Low cost canal sealing.* London. (Pamphlet)

28. PORTLAND CEMENT ASSOCIATION. *Irrigation with concrete pipe.* Chicago, Ill. 1952

29. Bottom-dumps ride travelling unloader, deliver 16 yd of concrete in 70 sec. 1968 *Construction Methods and Equipment,* 50.

30. PORTLAND CEMENT ASSOCIATION. *Soil-cement laboratory handbook.* See C21. 1959.

31. PORTLAND CEMENT ASSOCIATION. *Concrete pipe for irrigation.* Skokie, Ill. 1969

32. AMERICAN SOCIETY OF AGRICULTURAL ENGINEERS. *Engineering guidelines for the installation of flexible membrane linings. Proposed ASAE recommendations prepared by the ASAE Soil and Water Division Committee on Flexible Membrane Linings.* St. Joseph, Mich. SW-247.

33. ARMANIER, J. et al. *Les revêtements des canaux d'irrigation.* Arles. 34 p. 1953

34. ALEXEJEWSKI, E.E. The water resources policy of the Soviet Union. *Wasser* 1969 *und Boden,* 21(9).

35. BENSON, J.R. How to prevent lagoon seepage. *Public Works,* 43: 111-114. 1962

36. BARRAGAN, M.S. *Report on circular canals for large irrigation areas.* Paper 1969 presented at the seventh Congress of the International Commission on Irrigation and Drainage, Mexico.

37. BOOHER, L.J. *Surface irrigation.* Rome, FAO. FAO Agricultural Development
    1974   Paper No. 95. 160 p.

38. CORRY, J.A. & SCOTT, V.H. *Plastic film for lining irrigation ditches.* 11 p.
    (Unpublished)

39. DIXON SMITH, W. Canal and reservoir lining with asphalt. *Civil Engineering,*
    1962   32(5): 64-67.

40. DIRMEYER, R.D. & SHEN, R.T. *Sediment sealing of irrigation canals.* Fort
    1960   Collins, Colo., Colorado State University. 124 p.

41. DIRMEYER, R.D. *First quarterly progress report of bentonite sealing investiga-*
    1960   *tions, February-May 1960.* Fort Collins, Colo., Colorado State
    University, Experiment Station. 5 p.

42. DIRMEYER, R.D. *Third quarterly progress report of bentonite sealing investi-*
    1960   *gations, August-November 1960.* Fort Collins, Colo., Colorado
    State University, Experiment Station. 12 p.

43. DIRMEYER, R.D. et al. *Sealing rocky ditches.* Fort Collins, Colo., Colorado
    1962   State University, Cooperative Extension Service. Circular 203-A
    (revised). 17 p.

44. DIRMEYER, R.D. *Final report of clay sealing investigations for the period of*
    1961   *1 February 1960 to 31 January 1961.* Fort Collins, Colo., Colorado
    State University, Experiment Station. 10 p.

45. DELIGNE, B. Les travaux d'étaucheite du Canal du Nord. *Construction,*
    1964   7-8(19).

46. ELLSPERMAN, L.M. & HICKEY, M.E. *The use of asphalt in hydraulic con-*
    struction by the Bureau of Reclamation 1946-1959. Paper presented at the third annual Kansas Asphalt Paving Conference,
    University of Kansas, Lawrence, Kansas.

47. FREVERT, R.K. et al. *Engineering in soil and water conservation.* Ann
    1953   Arbor, Mich., Edwards.

48. GIRSHKAN, S.A. *Some problems of designing irrigation canals.* In International Commission on Irrigation and Drainage. *Annual Bulletin*
    1952   *1952,* p. 34-38. Delhi.

49. HICKEY, M.E. *Laboratory and field studies of asphaltic materials for control-*
    1961   *ling canal seepage losses.* Paper presented at the Conference on
    the Use of Asphalt in Hydraulic Construction, Pacific Coast Division. Bakersfield, Calif., Asphalt Institute. 54 p.

50. HOLMAN, H. Polythene linings for irrigation ditches. *North Dakota Farm*
    1965   *Research,* 23(11): 11-16.

51. HUMPHREYS, A.S. & LAURITZEN, C.W. *Hydraulic and geometrical relation-*
    1964   *ships of lay-flat irrigation tubing.* Washington, D.C., U.S. Department of Agriculture in cooperation with Utah Agricultural Experiment Station. Technical Bulletin 1309. 38 p.

52. JONES, C.W. & LOWITZ, C.A. Compacted loessial-soil canal linings. *Journal*
    1962   *of the Irrigation and Drainage Division, American Society of Civil*
    Engineers, 88.

53. JOHNSON, E. *An evaluation of soil-cement for channel stabilization.* Paper
    1961   presented at the 1961 Annual Meeting of the American Society
    of Agricultural Engineers. 9 p.

54. JONES, C.W. & LOWITZ, C.A. *Compacted loessial-soil canal linings. Pro-*
1966 *ceedings of the American Society of Civil Engineers*, 88 (IR4, Pt 1): 1-22.

55. KOUKAL, M. Modernes Verfahren zur Auskleidung kleinerer Kanäle. *Wasser*
1970 *und Boden*, 22(3).

56. KASSIF, G. et al. *Failure mechanism of canal lining in expansive clay.* Proc.
1967 ASCE. (SM1) No. 5068.

57. KRISHNA, M.B. & YENGOR, I. Concrete lining of high level Tungabhora
1967 Project. *Indian Concrete Journal*, 41(3).

58. LAURITZEN, C.W. *Lining irrigation laterals and farm ditches.* Washington,
1961 D.C., U.S. Department of Agriculture. Agriculture Information Bulletin No. 242. 11 p.

59. LAURITZEN, C.W. *Flexible membranes for water application.* Pamphlet.
1968 Reprinted from April 1968 issue of *World Irrigation*. 7 p.

60. LAURITZEN, C.W. *Butyl for the collection, storage and conveyance of water.*
1967 Logan, Utah Agricultural Experiment Station. Bulletin 465.

61. LAURITZEN, C.W. & GRIFFIN, R.E. *Plans for concrete slipforms.* Logan, Utah State University, Extension Services.

62. LAURITZEN, C.W. Soil cement linings for canals. *Utah Science*, 30(1): 10-14.
1969

63. LOVAS, L. *Canal construction and lining methods in the Chambal commanded area.* FAO/UNDP/SF Project IND 60.

64. MCNAUNCE, M.A. *Sealing irrigation ditches and canals with bentonite sedi-*
1960 *mentation.* Paper No. 60-701 presented at the 1960 Winter Meeting of the American Society of Agricultural Engineers, Memphis, Tennessee.

65. MCWHORTER, J.C., CARPENTER, T.G. & CLARK, R.N. Artificial channel liners.
1969 *Agricultural Engineering*, 50: 514-515, 518.

66. MYERS, L.E. et al. Sprayed asphalt pavements for water harvesting. *Journal*
1967 *of the American Society of Civil Engineers*, 93(IR3): 79-97.

67. NAZIR AHMAD & ABDUR RAZZAQ. *A stable and impervious lining for canals*
1960 *of West Pakistan.* West Pakistan Engineering Congress, Lahore, Pakistan. Paper No. 344. 56 p.

68. PILLSBURY, A. *Concrete pipe for irrigation.* Berkeley, Calif., California Agricultural Experiment Station, Extension Service. Circular 418.

69. QUACKENBUSH, T.H. How to install flexible membrane canal linings. *Agri-*
1967 *cultural Engineering*, 48: 500-501.

70. REGINATO, R.J. & MYERS, L.E. *Repair of cracks in concrete channel linings.*
1966 Washington, D.C., U.S. Agricultural Research Service. ARS-41-121. 8 p.

71. REGINATO, R.J. et al. *Sodium carbonate for reducing seepage from ponds.*
1968 Phoenix, Ariz., U.S. Water Conservation Laboratory. 6 p.

72. ROGERS, E.H. Soil cement linings for water containing structures. *In* Black,
1965 C.A., ed. *Methods of soil analysis*, p. 94-105. Madison, Wis., American Society of Agronomy. Agronomy Monograph No. 9.

73. RENEA, S. The use of plastic sheets for waterproofing of irrigation canals.
    1966   *Hidrotehnica, Gospodarirea Apelor, Meteorologia*, 11(4).

74. ROBINSON, E.P. *Watertight structures for farm channels*. Melbourne, Victoria
    1964   State Rivers and Water Supply Commission.

75. STAFF, C.E. Seepage prevention with impermeable membranes. *Civil Engi-*
    1967   *neering*, 37(2): 44-46.

76. STAMATI, P.P. *International Seminar on Major Irrigation Canals and their
    Equipment. Construction of major irrigation canals.* Nicosia,
    Cyprus Water Development Department. 10 p.

77. SCOTT, V.H. Small ditch seepage controlled. *California Agriculture*, 9(11):
    1955   9, 14.

78. SHEN, R.T. *Sealing sandy ditches with the bentonite dispersion method.*
    1959   Fort Collins, Colo., Colorado State University. Circular 202-A. 9 p.

79. SHEN, R.T. *Mixing bentonite for sealing purposes.* Fort Collins, Colo.,
    1959   Colorado State University. Circular 204-A. 9 p.

80. SMITH, C. *Installing plastic lining in McCaskey lateral.* Reprint from
    Reclamation Era. Washington, D.C., U.S. Bureau of Reclamation.
    3 p.

81. SEWELL, J.I. Pond sealing with chemicals in Tennessee. *Journal of Soil and*
    1969   *Water Conservation*, 24: 16-18.

82. SALLY, H.L. *Lining of earthen irrigation channels.* London, Asian Publishing
    1965   House.

83. SHEN, R.T. *Testing bentonite for sealing purposes.* Fort Collins, Colo.,
    1959   Colorado State University. Circular 205-A. 9 p.

84. TILP, P.J. *Capacity tests in large concrete lined canals.* Chicago, Ill. Portland
    1952   Cement Association.

85. UPPAL, H.L., MIDHA, D.C. & SOHAN SINGH. Evolving low-cost linings for
    1965   existing earthen channels: a combination type lining. *Indian Journal
    of Power and River Valley Development*, 15: 16-32.

86. UPPAL, H.L. *A new design of canal lining — one-tile lining.* (Abstract) 10 p.

87. VELEV, D. et al. *Lining of small irrigation channels with asbestos cement
    1960   flumes.* International Commission on Irrigation and Drainage,
    Fourth Congress, Madrid. Communication C21. 16 p.

88. WILSON, R.J. *Use of chemical soil sealants to reduce seepage from canals.*
    1962   Paper presented at a meeting of the American Society of Agri-
    cultural Engineers at Fort Collins, Colorado. 15 p.

89. WOODFORD, T.V.D. Linings for irrigation canals. *In* International Commission
    1953   on Irrigation and Drainage. *Annual Bulletin 1953*, p. 19-22, 37.

90. U.S. AGENCY FOR INTERNATIONAL DEVELOPMENT. *Design criteria, construction
    1970   guide and material standards for irrigation pipelines.* Report to
    the Government of India. Washington, D.C.

91. DONEEN, L.D. & TANJI, K.K. *Chemistry of lime in ammoniated waters for
    sealing concrete pipelines.* Berkeley, Calif., California Agricultural
    Experiment Station. Bulletin 841.

92. MICHAEL, A.M., REESER, R.M. & KNIERIM, G.C. *How to improve irrigation farming in India.* Jaipur, University of Udaipur. Extension Bulletin No. 2.
    1965

D. *Economics of lining*

1. ARMANIER, J. et al. *Les revêtements des canaux d'irrigation.* Arles. 34 p.
   1953

2. BAUDINI, A. Economic problem of irrigation canals: seepage losses. *Journal of the Irrigation and Drainage Division, Proceedings of the American Society of Civil Engineers,* JR4, December 1966. (Discussion, June 1967)
   1966

3. GRANT, E.L. Engineering economy in water resources planning. *In* Linsley, R.K. and Franzini, B., eds. *Water resources engineering,* p. 360-377. New York, McGraw-Hill.
   1964

4. GREENSHIELDS, E.L. *Report to the Government of Pakistan on some economic aspects of the Ganges-Kobadak Irrigation Development Scheme.* Rome, FAO. FAO/EPTA Report No. 1044.
   1959

5. JOHNSON, J.E. Figuring the dollar cost of canal linings. *Agricultural Engineering,* 49: 17, 24.
   1968

6. JOSEPH, A.P. Lining of irrigation channels — economic analysis. *Indian Farming,* 18: 41-43.
   1968

7. JOHNSON, J.E. *Economic considerations of canal linings.* Paper presented at the 1966 Winter Meeting of the American Society of Agricultural Engineers, Chicago, Ill. 9 p.
   1966

8. TIMMONS, F.L. & KLINGMAN, D.L. Control of aquatic and bank vegetation and phreatophytes. *In* American Association for the Advancement of Science. *Water and agriculture,* p. 157-170. Washington, D.C. Publication No. 62.
   1960

9. TIMMONS, F.L. *Weed control in western irrigation and drainage systems: losses caused by weeds; cost and benefits of weed control.* Washington, D.C., U.S. Department of Agriculture and U.S. Department of the Interior. 22 p.
   1960

10. PUNJAB. IRRIGATION AND POWER DEPARTMENT. *Lining of water courses in Bahawalpur irrigation region.* Punjab.
    1972

E. *Operation and maintenance*

1. U.S. BUREAU OF RECLAMATION. *Operation and maintenance equipment and procedures.* Denver, Colo. Release No. 40. 23 p.
   1962

2. U.S. BUREAU OF RECLAMATION. *Report on asphalt undersealing a portion of the concrete canal lining, Wyoming Canal-Riverton project.* Denver, Colo. Bituminous Laboratory Report No. B-15. 13 p.
   1950

3. Recent advances in maintenance of irrigation channels and drains. *In* International Commission on Irrigation and Drainage. *Annual Bulletin 1966,* p. 33-39.
   1966

4. BALCONI, R.B. Weeds — water robbers. *Journal of Soil and Water Conservation,* 5: 165-168.
   1950

5. CRAFTS, A.S. *Control of aquatic and ditch-bank weeds.* Berkeley, Calif.,
1949   California Agricultural Experiment Station, Extension Service. Circular 158. 15 p.

6. HILL, R.A. Operation and maintenance of irrigation systems. *Transactions*
1952   *of the American Society of Civil Engineers,* 117: 72-88.

7. SHAROV, I.A. *Summaries and extracts from selected chapters of Operation*
1958   *of hydraulic reclamation systems.* Rome, FAO. Informal Working Bulletin. 79 p.

F. *Miscellaneous*

1. ASPHALT INSTITUTE. *Asphalt in hydraulic structures.* New York. Manual
1961   Series No. 12 (MS-12). 128 p.

2. INDIA. PLANNING COMMISSION. COMMITTEE ON NATURAL RESOURCES. *Study*
1963   *on wastelands including saline alkali and waterlogged lands and their reclamation measures.* New Delhi. 237 p.

3. U.S. BUREAU OF RECLAMATION. *Earth manual: a guide to the use of soils as*
1960   *foundations and as construction materials for hydraulic structures.* Denver, Colo. 751 p.

4. U.S. BUREAU OF RECLAMATION. *Design of small dams.* Denver, Colo. 611 p.
1965

5. U.S. BUREAU OF RECLAMATION. *Concrete manual: a manual for the control of*
1966   *concrete construction.* 7th ed. Denver, Colo. 642 p.

6. KHANGAR, S.D. *Waterlogging in Western Punjab.* Punjab Engineering Con-
1947   gress. Paper No. 284.

7. PAZOS GIL, J. The improvement of irrigation systems. *In* European Com-
1968   mission on Agriculture. Working Party on Water Resources and Irrigation. *Report of the third session, Brussels, Belgium,* p. 121-142. Rome, FAO.

FAO SALES AGENTS AND BOOKSELLERS
AGENTS ET DÉPOSITAIRES DE LA FAO
LIBRERIAS Y AGENTES DE VENTAS DE LA FAO

Argentina	Editorial Hemisferio Sur S.R.L., Librería Agropecuaria, Pasteur 743, Buenos Aires.
Australia	Hunter Publications, 58A Gipps Street, Collingwood, Vic. 3066; The Assistant Director, Sales and Distribution, Australian Government Publishing Service, P.O. Box 84, Canberra, A.C.T. 2600, and Australian Government Publications and Inquiry Centres in Canberra, Melbourne, Sydney, Perth, Adelaide and Hobart.
Austria	Gerold & Co., Buchhandlung und Verlag, Graben 31, 1011 Vienna.
Bangladesh	Agricultural Development Agencies in Bangladesh, P.O. Box 5045, Dacca 5.
Barbados	Cloister Bookstore Ltd., Hincks Street, Bridgetown.
Belgique	Service des publications de la FAO, M.J. De Lannoy, rue du Trône 112, 1050 Bruxelles. CCP 000-0808993-13.
Bolivia	Los Amigos del Libro, Perú 3712, Casilla 450, Cochabamba; Mercado 1315, La Paz; René Moreno 26, Santa Cruz; Junín esq. 6 de Octubre, Oruro.
Brazil	Livraria Mestre Jou, Rua Guaipá 518, São Paulo 10; Rua Senador Dantas 19-S205/206, Rio de Janeiro.
Brunei	MPH Distributors Sdn. Bhd.,71/77 Stamford Road, Singapore 6 (Singapore).
Canada	Renouf Publishing Co. Ltd., 2182 Catherine St.West, Montreal, Que. H3H 1M7.
Chile	Biblioteca, FAO Oficina Regional para América Latina, Av. Providencia 871, Casilla 10095, Santiago.
China	China National Publications Import Corporation, P.O. Box 88, Peking.
Colombia	Litexsa Colombiana Ltda., Calle 55, N° 16-44, Apartado Aéreo 51340, Bogotá.
Costa Rica	Librería, Imprenta y Litografía Lehmann S.A., Apartado 10011, San José.
Cuba	Instituto del Libro, Calle 19 y 10, N° 1002, Vedado.
Cyprus	MAM, P.O. Box 1722, Nicosia.
Denmark	Ejnar Munksgaard, Norregade 6, Copenhagen S.
Ecuador	Su Librería Cía. Ltda., García Moreno 1172, Apartado 2556, Quito.
El Salvador	Librería Cultural Salvadoreña S.A., Avenida Morazán 113, Apartado Postal 2296, San Salvador.
España	Librería Mundi Prensa Libros S.A., Castelló 37, Madrid-1; Librería Agrícola, Fernando VI, 2, Madrid - 4.
Finland	Akateeminen Kirjakauppa, 1 Keskuskatu, Helsinki.
France	Editions A. Pedone, 13 rue Soufflot, 75005 Paris.
Germany, F.R.	Alexander Horn Internationale Buchhandlung, Spiegelgasse 9, Postfach 3340, Wiesbaden.
Ghana	Ghana Publishing Corporation, P.O. Box 3632, Accra.
Grèce	" Eleftheroudakis ", 4 Nikis Street, Athènes.
Guatemala	Distribuciones Culturales y Técnicas " Artemis ", Quinta Avenida 12-11, Zona 1, Guatemala.
Guyana	Guyana National Trading Corporation Ltd., 45-47 Water Street, Georgetown.
Haïti	Max Bouchereau, Librairie " A la Caravelle ", B.P. 111B, Port-au-Prince.
Honduras	Editorial Nuevo Continente S. de R.L., Avenida Cervantes 1230-A, Apartado Postal 380, Tegucigalpa.
Hong Kong	Swindon Book Co., 13-15 Lock Road, Kowloon.
Iceland	Snaebjörn Jónsson and Co. h.f., Hafnarstraeti 9, P.O. Box 1131, Reykjavik.
India	Oxford Book and Stationery Co., Scindia House, New Delhi; 17 Park Street, Calcutta.
Indonesia	P.T. Gunung Agung, 6 Kwitang, Djakarta.
Iran	Iran Book Co. Ltd.,127 Nadershah Avenue, P.O. Box 14-1532, Tehran; Economist Tehran, 99 Sevom Esfand Avenue, Tehran (sub-agent).
Iraq	National House for Publishing, Distributing and Advertising, Rashid Street, Baghdad.
Ireland	The Controller, Stationery Office, Dublin.
Israel	Emanuel Brown, P.O. Box 4101, 35 Allenby Road and Nachlat Benyamin Street, Tel Aviv; 9 Shlomzion Hamalka Street, Jerusalem.
Italie	Distribution and Sales Section, Food and Agriculture Organization of the United Nations, Via delle Terme di Caracalla, 00100 Rome; Libreria Scientifica Dott. L. De Biasio " Aeiou ", Via Meravigli 16, 20123 Milan; Libreria Commissionaria Sansoni " Licosa ", Via Lamarmora 45, C.P. 552, 50121 Florence.
Jamaica	Teachers Book Centre Ltd., 96 Church Street, Kingston.
Japan	Maruzen Company Ltd., P.O. Box 5050, Tokyo Central 100-31.
Kenya	The E.S.A. Bookshop, P.O. Box 30167, Nairobi.
Korea, Rep. of	The Eul-Yoo Publishing Co. Ltd., 5 2-Ka, Chong-ro, Seoul.